Scotland's leading educational publishers

D0313181

CfE Higher
CHEMISTRY
SUCCESS GUIDE

CfE Higher CHEMISTRY
SUCCESS GUIDE

6	7	8
C	N	O
Carbon	Nitrogen	Oxygen
12.011	14.007	15.999
14	15	16
Si	P	S
Silicon	Phosphorus	Sulfur
28.085	30.97376	32.06
32	33	34
Ge	As	Se
Germanium	Arsenic	Selenium
72.63	74.9216	78.96
50	51	52
Sn	Sb	Te
Tin	Antimony	Tellurium
118.71	121.76	127.6
82	83	84
Pb	Bi	Po
Lead	Bismuth	Polonium
207.2	208.9804	(209)
114	115	116
Fl	Uuq	Lv
	Ununpentium	Livermorium
(289)	(288)	(293)

Bob Wilson

001/230415

10 9 8 7 6 5 4 3 2 1

ISBN 9780007554393

Published by
Leckie & Leckie Ltd
An imprint of HarperCollins*Publishers*
Westerhill Road, Bishopbriggs, Glasgow, G64 2QT
T: 0844 576 8126   F: 0844 576 8131
leckieandleckie@harpercollins.co.uk   www.leckieandleckie.co.uk

Special thanks to
Lee Mulvey Haworth (project management); Helen Bleck (copy-edit); QBS (layout and illustration); Barry McBride (proofreading); Donna Cole (proofreading); Ink Tank (cover design)

A CIP Catalogue record for this book is available from the British Library.

Acknowledgements
Whilst every effort has been made to trace the copyright holders, in cases where this has been unsuccessful, or if any have inadvertently been overlooked, the Publishers would gladly receive any information enabling them to rectify any error or omission at the first opportunity.

Printed in Italy by Grafica Veneta S.P.A.

Unit 1: Chemical changes and structure

Unit 2: Nature's chemistry

Contents

Unit 3: Chemistry in society

Researching chemistry

The Higher Success Guide

About this Success Guide

The guide covers all of the mandatory content in the key areas of the Higher chemistry course.

The revision material is laid out over a double page spread (sometimes four pages) in a clear, concise way. The main headings and sub-headings make it clear which part of the course is being covered. There are 'Top Tips' throughout which emphasise important points and give hints on how to answer questions. Each section has a quick test to enable you to self assess so that you know your areas of strength and areas you need to look at in more detail. Answers to each test are at the back of the guide. There is also a glossary which gives you a quick and easy to find explanation of key words and phrases found in the book.

The Higher Chemistry Course

Structure
The CfE Higher Chemistry Course is made up of the following:
- Chemical changes and structure
- Nature's chemistry
- Chemistry in society

In order to achieve a pass at Higher, you need to pass all of the units as well as the course assessment.

Course assessment
This takes the form of a question paper (exam) and an assignment.

The exam carries a total of 100 marks and is made up of two sections:
- the objective test (20 marks) comprising 20 multiple-choice questions
- Paper 2 (80 marks) which is made up of a mixture of restricted and extended response questions.

The assignment is a written report, based on your own investigations and research into a topic in the course. It is worth 20 marks.

Preparation for both of these is critical to getting the best grade possible in your Higher Chemistry Course.

Controlling the rate of reaction 1

The importance of the rate of reaction

Generally speaking, industrial chemists want to speed up chemical reactions. They need to know how fast they can make a reaction go – typically, the faster the better. A reaction which is too slow is unlikely to be a commercial success. However, if the rate of reaction too rapid there is the risk of explosion.

The rate of chemical reactions are affected by:

- the concentration of dissolved reactants and the pressure in gas reactions;
- the size of solid reactant particles;
- the collision geometry – the angle at which reactants collide.

The effect of each of these factors can be explained by the collision theory.

Collision theory

TOP TIP

The more collisions that take place between reactant particles, the more chance there is of products being formed.

Simple **collision theory** states that for reactants to form products, they must first come in contact with each other (collide).

The reaction of marble chips (calcium carbonate) with hydrochloric acid can be used to illustrate some of the effects:

calcium carbonate + hydrochloric acid → calcium chloride + water + carbon dioxide

$$CaCO_3(s) \quad + \quad 2HCl(aq) \quad \rightarrow \quad CaCl_2(aq) \quad + H_2O(\ell) + \quad CO_2(g)$$

Concentration: the higher the concentration of the acid, the faster the reaction. The more particles there are, the more collisions there will be, and the greater the chance of reaction and products being formed. So, the higher the concentration, the more collisions and the faster the reaction.

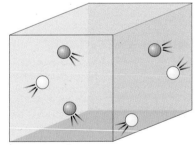

low concentration:
few particles in a given volume

Increasing the **pressure** in some reactions involving gases has a similar effect to increasing the concentration of the reactants. The gas molecules are compressed into a smaller space as the pressure is increased, so there is a greater chance of them colliding and going on to form products.

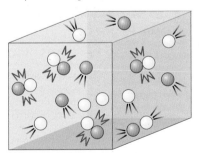

high concentration:
many particles in the same volume

Particle size: the smaller the particle size, the faster the reaction. When equal masses of small lumps and large lumps are reacted with the same concentration of acid, the small lumps react more quickly.

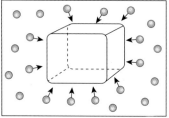 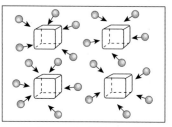

A large lump of solid has quite a small surface area. The reaction can only happen at the outside surface.

Smaller lumps of solid have a larger surface area at which the reaction can happen. The reaction rate is much faster.

Collision geometry

Not all collisions between reactants lead to products. One reason for this is the angle at which the molecules collide – the **collision geometry**. If two diatomic molecules collide end-on, this would not be the best alignment to form molecules of product. A better collision geometry would be if the molecules collided side-on or even end-to-side.

(a)

reactant molecules approach each other

ineffective collision – no reaction

reactants separate – no product

(b)

reactant molecules approach each other

effective collision – molecules have correct geometry

product molecules formed

Quick Test 1

1. Explain clearly the following observations:

   (a) increasing the concentration of nitric acid increases the speed at which it reacts with zinc metal;

   (b) powdered zinc reacts faster than lumps of zinc with nitric acid.

2. With the aid of diagrams, explain how collision geometry can affect the rate of a chemical reaction.

Controlling the rate of reaction 2

Temperature and activation energy

For many reactions, the higher the temperature, the faster the rate of reaction.

Acid warmed to 30°C will react much faster with marble chips than it would with acid at room temperature (approximately 20°C). Temperature is a measure of the average kinetic energy of the reactants. As the temperature increases, the kinetic energy of the reactants increases, i.e. they move faster. This results in more collisions taking place at a greater speed, so the chance of reactants forming products increases, and therefore the rate of reaction increases. However, a small rise in temperature (10°C) almost doubles the rate at which the acid reacts with the marble chips. This remarkable increase cannot be explained by an increase in the number of collisions, which is already high. The answer is that not only do collisions have to occur, but they have to have a minimum amount of kinetic energy, known as the **activation energy** (E_a). This is the minimum energy required to overcome the repulsion of the outer electrons.

The need for activation energy explains why the rate of some reactions is very slow unless energy is supplied:

- When natural gas in the laboratory mixes with the air, there is no visible sign that a reaction is occurring, even although many collisions are taking place. This is because so few molecules have the energy of activation. Sparking the gas/air mixture or holding a flame near it supplies the activation energy, so many more molecules can react. The activation energy for this reaction is high.

Energy distribution graphs

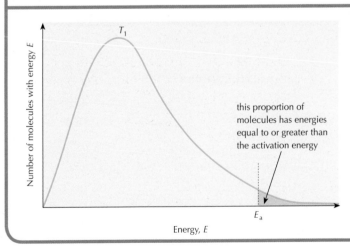

this proportion of molecules has energies equal to or greater than the activation energy

E_a

Energy, E

Number of molecules with energy E

T_1

Temperature can be considered to be a measure of the average kinetic energy of the particles in a substance. At a given temperature some particles will have high kinetic energy, and others low kinetic energy. Most particles will have kinetic energy somewhere in between. This can be displayed in an **energy distribution diagram,** as is shown opposite.

The total area under the graph shows the energy distribution of all the reactants at temperature T_1. Only a relatively small number of reactant particles have energy greater than the activation energy, E_a. This is shown as the shaded area under the graph.

If the temperature is increased, although there are the same number of particles, their energy distribution is different. More particles have higher kinetic energy. This is shown in the graph below, where the energy distribution at T_2 (higher than T_1) is superimposed on the diagram for temperature T_1.

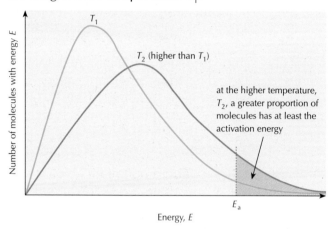

at the higher temperature, T_2, a greater proportion of molecules has at least the activation energy

Notice that the shape of the graph at T_2 is distorted slightly to the right, to reflect that more particles have higher energy. The height of the graph is also lower than the graph at T_1. The shaded area under the graph shows that there are more than twice as many reactant particles with energy greater than E_a, so more than twice as many particles now have enough energy to react when they collide. This means that for a relatively small increase in temperature, there is a big increase in the rate of the reaction.

TOP TIP

A relatively small increase in temperature has a big effect on the rate of a reaction, because many more reacting molecules have energy equal to or greater than E_a.

Quick Test 2

1. In terms of activation energy, explain fully the reasoning behind advising people not to switch on lights or other electrical devices if they detect a gas leak.

2. The energy distribution diagram is for a reaction carried out at 40°C.

    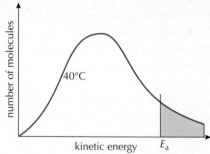

    (a) Add a line to the graph to show the energy distribution of the reactants, if the reaction were carried out at 30°C.

    (b) Explain what the shaded area under the graph represents.

3. A student wrongly stated, 'The rate of a reaction increases when the temperature increases, because the activation energy increases, which increases the number of collisions.' Give the correct explanation.

Controlling the rate of reaction 3

Relative rate

The relationship between concentration and rate can be studied in more detail by carrying out a 'clock reaction'. This is where the effect of changing the concentration of one of the reactants in a reaction can be measured by timing how long it takes for a reaction to reach a certain point – usually indicated by a colour change. Because the concentration of only one reactant is being changed, the time it takes to reach the same point in the reaction must be due to the one reactant. The longer it takes to reach the same point in the reaction, the slower the rate. The rate of the reaction is inversely proportional to time (t), so can be taken as the reciprocal of time, i.e. 1/t and is known as the relative rate (R).

A closer look at the effect of concentration on rate

A suitable clock reaction is hydrogen peroxide (H_2O_2) reacting with potassium iodide (KI) to form iodine (I_2).

$$H_2O_2(aq) + 2H^+(aq) + 2I^-(aq) \rightarrow 2H_2O(\ell) + I_2(aq)$$

The course of this reaction can be followed by carrying it out in the presence of small quantities of starch and thiosulphate ($S_2O_3^{2-}$) solutions. As the iodine molecules are produced, they immediately react with the thiosulphate ions and are converted back to colourless iodide ions:

$$I_2(aq) + 2S_2O_3^{2-}(aq) \rightarrow 2I^-(aq) + S_4O_6^{2-}(aq)$$

Once the thiosulphate ions have been used up, a blue/black colour suddenly appears, because the iodine molecules now have the opportunity to react with the starch. The time taken for the colour to first appear is noted.

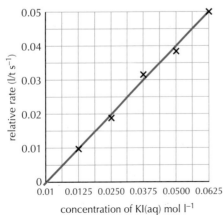

The graph is a straight line showing that rate is directly proportional to concentration, that is, if the concentration doubles, the rate doubles. This is the case for many chemical reactions.

A closer look at the effect of temperature on rate

The relationship between temperature and rate can be studied in more detail by studying the clock reaction between oxalic acid and acidified potassium permanganate:

$$5(COOH)_2(aq) + 6H^+(aq) + 2MnO_4^-(aq) \rightarrow 2Mn^{2+}(aq) + 10CO_2(g) + 8H_2O(\ell)$$

                           purple   colourless

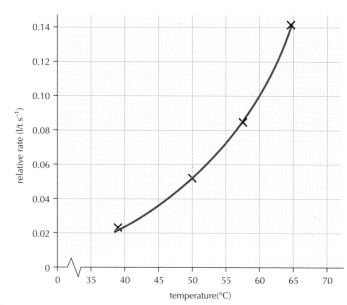

A series of experiments are carried out, in which only the temperature of the reaction mixtures was changed.

The graph, obtained from the results of such an experiment, shows that the rate is not directly proportional to temperature. A small rise in temperature results in a relatively large increase in rate. This supports the idea of activation energy.

Quick Test 3

1. (a) The reaction between hydrogen peroxide and potassium iodide was carried out using $0{\cdot}190$ mol l^{-1} iodide solution. A sharp colour-change was noted after 67 s.

    Calculate the relative rate of the reaction.

    (b) (i) From the graph of relative rate vs concentration, predict the relative rate when the concentration of the potassium iodide solution used is $0{\cdot}125$ mol l^{-1}.

    (ii) Justify your answer.

2. Look at the graph of relative rate against concentration.

    Calculate how long it would take before a colour-change to blue/black would be seen, if the concentration of the iodide solution was $0{\cdot}0318$ mol l^{-1}.

3. The reaction between oxalic acid and acidified permanganate solution was carried out at 25°C. The purple colour of the permanganate ion disappeared after 102 s.

    Calculate the relative rate of the reaction.

Controlling the rate of reaction 4

Interpreting rate graphs

The course of a reaction can be followed by measuring the volume of gas given off over time, and drawing a graph of volume of gas against time.

Experiment 1

A suitable experiment would be to observe the reaction of 40 cm³ of 0·2 mol l⁻¹ hydrochloric acid with excess magnesium ribbon at room temperature (20°C):

magnesium + hydrochloric acid → magnesium chloride + hydrogen

$$Mg(s) \quad + \quad 2HCl(aq) \quad \rightarrow \quad MgCl_2(aq) \quad + \quad H_2(g)$$

Interpreting the shape of the graph

At 1: The steepest part of the graph – the rate is at its highest. The maximum number of reactant particles are present, and the number of effective collisions per second is at its highest.

At 2: The graph is not so steep – the rate of reaction is slowing. Some of the reacting particles have been used up so the number of effective collisions per second is less.

At 3: The graph is horizontal – the reaction has stopped. The acid has been completely used up.

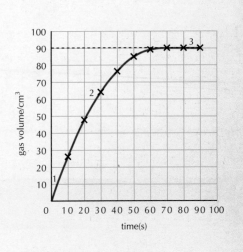

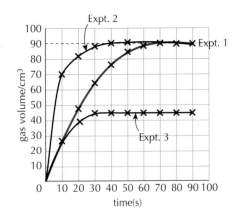

The graphs opposite are for the same reaction, but with some of the variables changed (quantities of reactant, particle size and temperature.) Expt. 1 (red line) are the original results.

TOP TIP

The slope of a graph indicates how fast the rate of reaction is. The steeper the slope, the faster the reaction.

Experiment 2

The reaction of 40 cm^3 of 0·2 mol l^{-1} hydrochloric acid with excess magnesium ribbon, but at a higher temperature than used in Experiment 1.

Interpreting the graph for Experiment 2

The graph is steeper than Experiment 1 initially, indicating that the reaction is faster. This is because the reaction is taking place at a higher temperature. More particles have energy greater than the activation energy. All the other variables are as in Experiment 1. The final volume of gas produced is the same as in Experiment 1, although it is produced faster in Experiment 2, due to the mass of magnesium ribbon and the number of moles of acid reacting is the same.

Experiment 3

The reaction of 20 cm^3 of 0·2 mol l^{-1} hydrochloric acid with excess magnesium ribbon, at the same temperature used in Experiment 1.

Interpreting the graph for Experiment 3

The graph has the same slope as Experiment 1 initially, indicating that the rate of reaction is the same. This is because the concentration of the hydrochloric acid, the temperature, and the amount of magnesium ribbon are the same in both experiments. The final volume of gas obtained is half of that in Experiment 1 because the number of moles of acid reacting is half that of the original.

Quick Test 4

1. Experiment 1 above was repeated using excess magnesium powder.

    (a) Sketch the graph for Experiment 1, and add a line to the graph for the experiment carried out with powdered magnesium.

    (b) Explain the shape of the graph you have drawn.

Controlling the rate of reaction 5

Reaction profiles

When chemical reactions occur, they are often accompanied by a significant change in energy. This is known as the **enthalpy change** and is given the symbol **ΔH**. ΔH is measured in **kJ** or **kJ mol^{-1}** when one mole of substance is reacted or produced.

When energy is released into the surrounding area, (most commonly in the form of heat) the reaction is said to be **exothermic** and the value is given a negative sign. The energy released when new products are formed in an exothermic reaction must have come from the reactants, which have potential energy (PE). Some of this potential energy is released as heat during the reaction. Combustion and neutralisation are exothermic reactions, and so have negative ΔH values.

When energy is taken in from the surrounding area, the reaction is said to be **endothermic** and the value is given a positive sign.

Enthalpy changes can be shown in potential energy diagrams – the activation energy (E_a) is also shown in the diagrams.

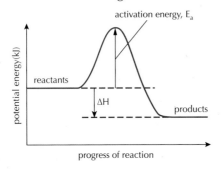
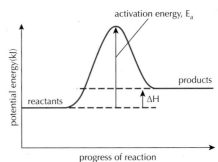

If potential energy values are added to the diagrams, the enthalpy change (ΔH) for a reaction can be calculated by subtracting the enthalpy of products from the enthalpy of the reactants:

$$\Delta H = H_{(products)} - H_{(reactants)}$$

The activation energy can be calculated by subtracting the enthalpy of the reactants from the energy maximum in the graph.

TOP TIP

In exothermic reactions, the PE of the products is less than the PE of the reactants. The 'missing' PE is given out as heat into the surrounding area. In endothermic reactions the PE of the products is greater than the PE of the reactants. The 'extra' PE is taken in from the surrounding area.

TOP TIP

Exothermic reactions always have a negative sign in front of the ΔH value to indicate the reaction is exothermic. Endothermic reactions should have a positive sign in front of the ΔH value – it is often missed out though.

Example

Calculate (i) the enthalpy change, and (ii) the activation energy from the potential energy diagram shown.

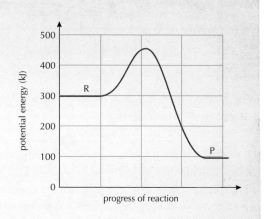

Worked answer

(i) $\Delta H = H_{(products)} - H_{(reactants)}$

      $= 100 - 300$

      $\Delta H = -200$ kJ

(ii) $E_a = 450 - 300$

       $E_a = 150$ kJ

The potential energy diagrams show the activation energy as an energy barrier, which the reactants have to overcome in order to form products. Reactant molecules, with energy equal to or greater than the activation energy, can go on to form products. Reactant molecules with energy lower than the activation energy will not go on to form products.

The lower the activation energy, the faster the rate of reaction.

The higher the activation energy, the slower the rate of reaction.

This explains why although petrol is very flammable, you can safely fill up a car's tank at the pump, because the activation energy needed for the combustion of petrol is high. It also explains why there are 'No Smoking' signs on display at petrol stations. A flame near petrol would supply the activation energy needed for combustion to occur.

Quick Test 5

1.   (a) Calculate (i) the enthalpy change, and
         (ii) the activation energy from the potential energy
         diagram shown.
     (b) (i)   Is the reaction exothermic or endothermic?
         (ii) Justify your answer.

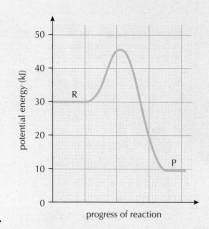

2.   Reactions (a) and (b) are associated with iron production:
     (a) $C(s) + O_2(g) \rightarrow CO_2(g)$ $\Delta H = -394$ kJ mol^{-1}
     (b) $C(s) + CO_2(g) \rightarrow 2CO(g)$ $\Delta H = +173$ kJ mol^{-1}

         (i)   Sketch potential energy diagrams for both
               reactions (they do not have to be drawn to scale).
         (ii) Indicate the ΔH and E_a on each diagram.
         (iii) For each reaction, state whether it is exothermic
               or endothermic. Justify your choices.

15

Controlling the rate of reaction 6

The activated complex

As a reaction progresses there is a stage reached where an intermediate product is formed known as the **activated complex**. The top of the potential energy barrier represents the point at which the activated complex has formed.

The activation energy can be defined as the minimum energy needed by colliding particles to form the activated complex. Activated complexes are very unstable, and only exist for a short time. Not all activated complexes lose energy by going on to form products. They can break down to reform reactants. This is indicated by the arrows in the potential energy diagram.

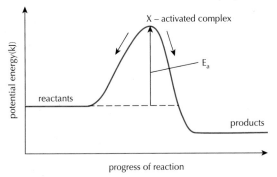

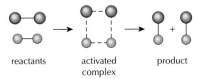

The structure of many activated complexes is not known. When two diatomic molecules react, a 'square' activated complex can form – the dotted lines indicate bonds breaking and bonds forming.

Catalysts and activation energy

A catalyst is a substance which increases the rate of a particular reaction, without being used up in the reaction.

Catalysts are essential in many important industrial processes, some of which were covered in the National 5 course – these are summarised in the table below:

Process	Catalyst	Reaction
Haber – making ammonia	iron	$N_2 + 3H_2 \rightarrow 2NH_3$
Ostwald – making nitric acid	platinum	$4NH_3 + 5O_2 \rightarrow 4NO + 6H_2O$
Hydrogenation – making margarine	nickel	Unsaturated oils + $H_2 \rightarrow$ saturated fats

The catalyst acts as a site for the reaction to take place. If no catalyst were used, the reaction would take place via a different route and at a different rate. The steps in a catalyst-assisted pathway each have lower activation energy than the route without the catalyst, so more molecules have enough energy to react. This is shown in the potential energy diagram on page 17.

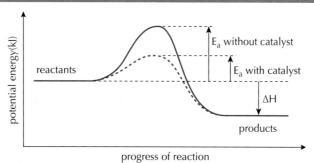

Potential energy diagram comparing the Ea of a catalysed reaction with the uncatalysed reaction

The catalysts in the table on page 16 work because the reactant molecules attach themselves temporarily to the surface of the catalyst. This is known as adsorption.

Whenever a catalyst is used, its surface area is made as big as possible to provide a large area for collisions between the catalyst and reactant molecules. The catalyst is often referred to as 'finely divided'.

TOP TIP

Don't get mixed up between adsorption and absorption. Adsorption is when reactants attach to the surface of a catalyst. A piece of kitchen roll soaking up spilled water is an example of absorption.

Quick Test 6

1. (a) An activated complex is formed during many reactions. Explain what is meant by 'activated complex'.

    (b) Two covalentlty bonded molecules, A-A and B-B, react to form a product, A-B.

    $A_2 + B_2 \rightarrow 2AB$

    Sketch the shape of a possible activated complex.

2. Catalytic converters in car exhaust systems use platinum to remove harmful gases.

    (a) Name the process which takes place at the surface of the catalyst.

    (b) Describe the process which takes place at the surface of the catalyst.

    (c) Explain why more molecules are able to react when a catalyst is used in a chemical reaction.

Periodicity 1

The first 20 elements

Elements on the periodic table are arranged in order of increasing atomic number. This allows chemists to make accurate predictions of physical properties and chemical behaviour for any element, based on its position in the periodic table. The similarities between the elements within a group are explained in terms of the number of electrons in the outer energy level (shell). This similarity in the electron arrangement leads to a periodicity (repeating pattern) in the properties of elements, so that the pattern of properties shown by the elements in one period is repeated in the next period.

The first 20 elements can be used to illustrate periodicity. They can be roughly divided into four areas where the bonding and structure within each area are similar.

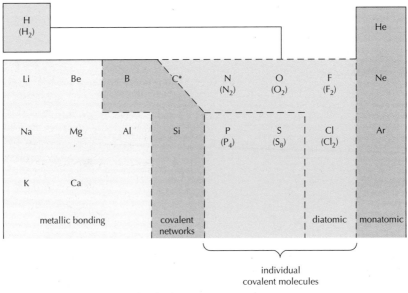

* Carbon can exist as networks and molecules

Monatomic elements – The noble gases (group 0)

At the far right of the periodic table we find the noble gases in group 0. Both their group name and number indicate a lack of reactivity. They are **monatomic** – they exist as individual atoms.

Each noble gas atom has a full outer energy level, and so does not need to react to achieve this. However, they are liquified by lowering the temperature, which indicates there must be some force of attraction between the atoms, which is significant at a lower temperature but not at room temperature. These forces of attraction are known as **London dispersion forces**. They are caused by the continual movement of electrons in an atom, which causes a temporary uneven distribution of charge at opposite sides of an atom. This is known as a temporary dipole. This means that one side of the atom is temporarily slightly negative (δ^-), resulting in the other side being temporarily slightly positive (δ^+). This in turn induces a temporary dipole in a neighbouring atom. This results in the δ^- side of one atom attracting the δ^+ side of a neighbouring atom, so a force of attraction is formed between them.

> **TOP TIP**
>
> London dispersion forces exist in molecular substances, not only between the atoms of noble gas elements.

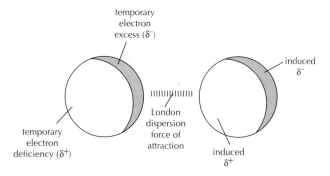

Diatomic elements
H_2, N_2, O_2, F_2 and Cl_2

> **TOP TIP**
>
> It is important to realise that when a molecular substance melts or boils, it is the weak London dispersion forces between molecules (**intermolecular forces**) which are broken, and not the strong covalent bonds holding the atoms together in the molecule (**intramolecular forces**).

Diatomic elements exist as two atoms covalently bonded.

Most are gases at room temperature that are easily liquified at low temperatures. Bromine is a liquid, and iodine a solid at room temperature. This indicates that there are forces of attraction between the molecules. London dispersion forces attract the molecules to each other. However, in some of the molecules, these forces are so small that there is enough energy at room temperature to overcome the weak force of attraction, and separate the molecules from each other. This is why most of the diatomic molecules are gases. Bromine and iodine are bigger molecules, and have more electrons than the smaller gas molecules. Therefore, the London dispersion forces are much stronger between their molecules, and it takes more energy to separate them. This explains why they exist as a liquid and a solid at room temperature.

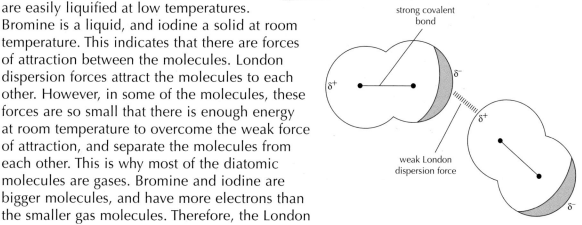

Quick Test 7

1. A student writing about noble gases stated: 'Helium is a gas which exists as individual atoms with weak forces of attraction between the atoms.' Describe fully the force of attraction acting between the atoms of noble gases.

2. (a) Describe the trend in melting points and boiling points of the halogens as you go down group 7 of the periodic table.

    (b) Explain the trend you described in (a).

Periodicity 2

Bigger molecules: P_4, S_8 and fullerenes (C)

Phosphorus and **sulfur** exist as solids with low melting and boiling points. Phosphorus exists as P_4 molecules, and sulfur as S_8 molecules. The atoms which make up the molecules are held together by covalent bonds, and the molecules are held together by London dispersion forces. Due to the size of the molecules, there are more electrons than there are in, for example, diatomic molecules.

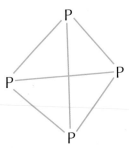

P_4: mp = 44°C
bp = 280°C

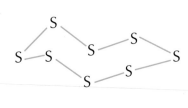

S_8: mp = 113°C
bp = 445°C

Therefore the London dispersion forces are stronger. This means at room temperature, there is not enough energy to break the forces and separate the molecules, so the elements are solid.

Carbon had, for a long time, been thought only to exist in the form of graphite and diamond, which are covalent networks (see page 21). However, in the 1980s, a molecular form of carbon was discovered. The molecule had 60 carbon atoms (C_{60}) and was named Buckminsterfullerene, often shortened to **fullerene**. Subsequent larger fullerene molecules have been made. Fullerenes are comprised of a series of carbons arranged in hexagons and pentagons. These are joined to make a ball shape. Fullerenes can also exist as tube shapes, often called nanotubes.

Like all covalent molecular elements, the molecules in fullerenes are held together by London dispersion forces. Because the molecules are so large, a lot of energy is required to separate them. Fullerene sublimes (changes from solid to gas) at around 600°C.

Models of ball-shaped fullerenes.

Covalent network elements: B, C (as diamond and graphite) and Si

Some non-metal elements have extremely high melting and boiling points. This indicates that there must be forces other than London dispersion forces holding their structures together. This can be seen in the elements **carbon, silicon** and **boron**.

Diamond and graphite (C)

Until molecules of fullerene (C_{60}) were discovered in the 1980s, it was thought that carbon only existed as giant **covalent networks**, in the form of diamond and graphite. Diamond and graphite are formed naturally by the combined action of heat and pressure deep under the Earth's crust.

In **diamond**, each carbon atom is covalently bonded to four other carbon atoms in a tetrahedral arrangement. All the outer electrons in the atom of each carbon are used to make single covalent bonds with neighbouring atoms. This results in a giant covalent network. There are no individual molecules in a network, nor are there unbonded (delocalised) electrons in the diamond structure. Therefore the diamond is a non-conductor of electricity.

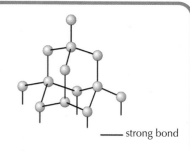

—— strong bond

The structure of diamond.

The covalent network structure of **graphite** shows that any one carbon atom is bonded to only three other carbon atoms, unlike diamond where it is four. The diagram of the structure of graphite shows that the atoms form hexagonal plates, which are held together by weak London dispersion forces. These layers are able to slide over each other, so powdered graphite can be used as a lubricant.

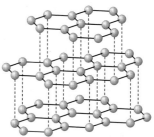

------ weak attractive forces

Because each carbon atom only uses three of its outer electrons for bonding, the fourth electron is delocalised. It can move through the structure in a similar fashion to the delocalised electrons in metals. This explains why graphite is a good conductor of electricity.

The internal structure of graphite, showing carbon atoms joined in layered hexagonal plates, which are held together by London dispersion forces.

Melting and boiling points

In order to separate the atoms in a covalent network, the strong covalent bonds have to be broken. This requires a lot of energy, and explains why elements with covalent network structures have high melting and boiling points.

Silicon (Si) and boron (B)

Silicon is in group 4 of the periodic table. It has similar properties to diamond, in that it has a very high melting point, is extremely hard, and in its pure state is a very poor conductor of electricity. Silicon forms four covalent bonds with four other silicon atoms, and forms a covalent network similar to diamond.

Boron is similar to carbon in its ability to form stable covalent networks. Boron is a very hard, black material with a high melting point of above 2000°C.

> **TOP TIP**
>
> Diamond and graphite are unusual, in that rather than melting when heated to very high temperatures, they sublime. Sublimation is the transition of a substance directly from the solid to the gas phase, without passing through an intermediate liquid phase.

Quick Test 8

1. Explain why sulfur has a higher melting point than phosphorus.

2. Fullerenes are molecular forms of carbon which exist as solids that sublime (solid → gas) at temperatures approximately 600°C and above. Describe the intramolecular and intermolecular forces present in fullerenes, and state why their sublimation temperatures are so high.

3. Explain fully the following observations:

   (a) Diamond's sublimation point is much higher than sulfur's boiling point.

   (b) Graphite conducts electricity but diamond doesn't.

   (c) Silicon has many properties similar to diamond.

Periodicity 3

Covalent radius

The **covalent radius** can be used as a measure of the size of an atom.

The covalent radius is taken as half the distance between the nuclei of two atoms, joined by a single covalent bond.

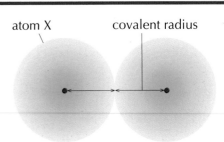

Change in covalent radius

Covalent radius decreases across a period →						
Li 134	Be 129	B 90	C 77	N 75	O 73	F 71

Covalent radius increases down a group ↓

Na 154	Going **down a group**, there is an increase in atomic number, i.e. the number of protons in the nucleus. This is also means that the number of electrons in each atom increases. The number of electron shells occupied increases, and the inner electron shells shield the outermost electrons from the full pull of the positive nucleus. This means the outer electrons can move further from the nucleus.
K 196	
Rb 216	Going **across a period**, the atomic number increases by one each time, so the number of protons in the nucleus increases by one. At the same time, each atom gains an electron but in the same shell. There is no increased shielding effect from the inner shells of electrons. Thus, as the atomic number increases across a period, the electrons in the outer shell are more strongly attracted to the nucleus, and so the covalent radius decreases.
Cs 235	

Ionisation energy

The **ionisation energy** is the energy required to remove one mole of electrons from one mole of atoms in the gaseous state. Ionisation energy is measured in kJ mol^{-1}. If one mole of electrons is removed, the energy required is known as the **first** ionisation energy. For example, the first ionisation energy for potassium would be written as:

$$K(g) \rightarrow K^+(g) + e^-$$

Change in ionisation energy

First ionisation energy generally increases across a period						
Li 520	Be 900	B 801	C 1086	N 1402	O 1314	F 1681

First ionisation energy generally decreases as you go down a group

Na 496	As you go **down a group**, the covalent radius of the atoms increases, i.e. the atoms get bigger. This means that the outer electrons are further away from the pull of the nucleus. Additionally, the number of filled electron energy levels increases, and they shield the outer electrons from the pulling effect of the nucleus. A combination of both effects makes it easier to remove an outer electron from an atom as you go down a group.
K 419	
Rb 403	As you go **across a period**, although electrons are being added, it is to the same energy level. The nuclear charge increases and the electrons are held more tightly, so it becomes harder to remove an outer electron from an atom.
Cs 316	

It is possible to remove two or more electrons from an atom. The second ionisation energy is the amount of energy required to remove a second mole of electrons.

Using calcium as an example:

First ionisation energy: $Ca(g) \rightarrow Ca^+(g) + e^-$ $\Delta H = +590$ kJ mol^{-1}

Second ionisation energy: $Ca^+(g) \rightarrow Ca^{2+}(g) + e^-$ $\Delta H = +1145$ kJ mol^{-1}

Second ionisation energies are always higher than first, because although electrons are being removed, the number of protons in the nucleus remains the same. Therefore, the pull on the remaining electrons is increased.

The total amount of energy required to remove two moles of electrons is the first ionisation energy and second ionisation energy added together, i.e. +1735 kJ mol^{-1} for Ca.

TOP TIP

In energy terms, it is not possible for atoms of group 2 elements to form a 3+ ion. This is because it would mean removing a third electron from an energy level closer to the nucleus, which requires too much energy.

Electronegativity

Electronegativity is the attraction an atom involved in a bond has for the electrons of the bond. Scientist Linus Pauling devised a scale, assigning numbers to elements based upon how electronegative that element is.

- Fluorine is the most electronegative element, and has a value of 4.0.
- Francium is the least electronegative, and has a value of 0.8.

The difference in electronegativity values of bonded atoms gives a good indication of the type of bonding that primarily exists between them.

Change in electronegativity

Electronegativity values generally decrease — down a group	Electronegativity values increase across a period ⟶						
	Li 1.0	Be 1.5	B 2.0	C 2.5	N 3.0	O 3.5	F 4.0
	Na 0.9	As you go **down a group**, the covalent radius increases and the outer electrons get further from the nucleus. The number of shielding shells of electrons also increases. These factors combine and result in the nucleus having less of an attraction for the bonding electrons.						
	K 0.8	As you go **across a period**, the covalent radius decreases and the number of shielding shells of electrons stays the same, so the attraction the nucleus has for the bonding electrons increases.						
	Rb 0.8							
	Cs 0.8							

TOP TIP

Covalent radii, ionisation energies and electronegativity values can all be found in the SQA data booklet.

Quick Test 9

Word bank

covalent, decreases, group, increases, nucleus, period, radius, shielding, stronger

1. Complete the following paragraph, using the word bank to help you.
    As you go down a (i) _____ in the periodic table, the covalent
    (ii) _____ increases. This is because the number of shielding
    shells of electrons (iii) _____ and the attraction between the
    (iv) _____ and the outermost electrons (v) _____.
    Going across a (vi) _____ the (vii) _____ radius decreases.
    This is because although the number of electrons increases, the number of
    (viii) _____ shells of electrons is the same. This means the attraction
    of the nucleus for the outermost electrons gets (ix) _____.

2. This table shows ionisation energy values (in kJ mol^{-1}) for a mole of magnesium
    atom forming ions:

first	second	third
738	1451	7733

    (a) Write equations to represent the first and second ionisation energies of magnesium.

    (b) Explain why there is an increase in the ionisation energy from first to second.

    (c) Explain why there is such a large difference between the second and third
        ionisation energy values.

    (d) Calculate the total energy required for a mole of gaseous magnesium atoms to form
        a mole of magnesium 2+ ions.

Word bank

across, covalent, decreases, down, electrons, nucleus, radius, shielding, stronger

3. Complete the following paragraph, using the word bank to help you.
    As you go (i) _____ a group in the periodic table, electronegativity
    (ii) _____. This is because the covalent (iii) _____ and
    number of shielding shells of (iv) _____ increase, and the attraction
    between the (v) _____ and the outermost electrons decreases.
    Going (vi) _____ a period the (vii) _____ radius decreases.
    This is because although the number of electrons increases, the number of
    (viii) _____ shells of electrons is the same. This means the attraction
    of the nucleus for the outermost electrons gets (ix) _____.

Structure and bonding 1

Types of bonding

Covalent – pure and polar

Covalent bonds are formed when non-metal atoms share pairs of electrons. When two atoms share bonding electrons equally, it is known as **pure covalent** bonding. This is because they have the same attraction for the bonding electrons – both atoms have the same **electronegativity** value (see page 24).

The bonds between chlorine atoms in a chlorine molecule (Cl_2) and phosphorus and hydrogen atoms in phosphine (PH_3) are pure covalent bonds.

$$Cl \overset{\bullet}{\underset{x}{-\!\!\!-}} Cl \qquad P \overset{\bullet}{\underset{x}{-\!\!\!-}} H \qquad \text{x and } \bullet = \text{bonding electrons}$$

Electronegativity value: $3 \cdot 0 - 3 \cdot 0$ (Difference = 0) $\qquad$ $2 \cdot 2 - 2 \cdot 2$ (Difference = 0)

Most covalent compounds are formed between elements with different electronegativity values. This means that one atom has a greater attraction for the bonding electrons than the other. The atom with the greater attraction for the bonding electrons will have a slightly negative charge (δ^-), leaving the other atom with a slightly positive charge (δ^+). A **permanent dipole** is formed. Bonding in which there is a permanent dipole is known as **polar covalent** bonding (polar bonding). The H–Cl bond in hydrogen chloride and the O–Cl bond in dichlorine monoxide are examples of polar covalent bonds.

$$H^{\delta+} \overset{\bullet}{\underset{x}{-\!\!\!-}} Cl^{\delta-} \qquad O^{\delta-} \overset{\bullet}{\underset{x}{-\!\!\!-}} Cl^{\delta+} \qquad \text{x and } \bullet = \text{bonding electrons}$$

Electronegativity value: $2 \cdot 2 - 3 \cdot 0$ (Difference = $0 \cdot 8$) $\qquad$ $3 \cdot 5 - 3 \cdot 0$ (Difference = $0 \cdot 5$)

Not all polar bonds are the same strength. The difference in electronegativity values gives a measure of how polar a bond is. As a general rule, if the difference in electronegativity is less than 2 the bond is covalent. The nearer the difference is to 2, the more polar a bond is. The nearer the difference is to zero, the less polar the bond is. (See Bonding continuum on page 27.)

Fluorine is the most electronegative element, with a value of $4 \cdot 0$. When bonded with hydrogen with a value of $2 \cdot 1$, the difference in electronegativity is $1 \cdot 9$. This means that the bond is extremely polar.

Ionic bonding

Differences in electronegativity values greater than 2 indicate that the electron movement from the atom of the element with the lower electronegativity value, to the atom of the element with the greater electronegativity is complete, resulting in the formation of ions. Ionic compounds exist as giant lattices in which the oppositely charged ions attract each other. The electrostatic attraction between the positive ions and the negative ions is not in any particular direction, as is the case with covalent bonds. Sodium chloride and calcium chloride are ionic.

Both compounds have an electronegativity difference greater than 2·0 and so are ionic, but CsCl is slightly more ionic than NaCl because it has the bigger electronegativity difference between the elements.

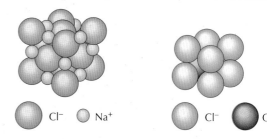

Cl⁻ ◯ Na⁺   Cl⁻ ⬤ Cs⁺

TOP TIP

Caesium is the least electronegative element and fluorine the most electronegative. Caesium fluoride is considered to be the most ionic compound because the difference in electronegativity is the highest possible (3·2).

TOP TIP

Beware of assuming all metal/non-metal compounds are ionic. For example, the properties of tin(IV) chloride show a lot of covalent character, which would be expected from the difference in electronegativity value between the two elements, which is 1·2.

The bonding continuum

The idea of a **bonding continuum** can be used to give a more realistic indication of the bonding in a compound. The bonding continuum has ionic bonding at one end and pure covalent bonding at the other end, with polar covalent bonding in between. The numbers in the diagram are the difference in electonegativity between the bonded atoms.

3.0	2.0	1.5	1.0	0.0
Li^+F^-	$Be^{2+}(F^-)_2$	$C^{\delta+}\!\!-\!\!F^{\delta-}$	$N^{\delta+}\!\!-\!\!F^{\delta-}$	$F\!\!-\!\!F$
ionic		polar covalent		pure covalent
X^+Y^-		$X^{\delta+}\!\!-\!\!Y^{\delta-}$		$X\!\!-\!\!Y$

TOP TIP

Although looking at the differences in electronegativity is a useful indicator of bonding, it has to be combined with chemical properties to be absolutely sure of the bonding present in a compound.

To say that bonding is on a continuum means that the type of bonding changes gradually as the difference in electronegativity between atoms increases. There is no sharp distinction between covalent and ionic bonds.

Quick Test 10

1. Use the electronegativity values in the SQA data booklet to predict the type of bonding which would predominate in the following compounds:

    (i) magnesium oxide; (ii) tin(iv) bromide; (iii) hydrogen iodide

    Justify each of your answers.

2. Explain the difference between pure covalent bonding and polar covalent bonding.

3. Explain what is meant by the bonding continuum.

Structure and bonding 2

Intermolecular forces

The **intermolecular forces** acting between molecules are known as **van der Waals forces**. There are three types of van der Waals force: **London dispersion forces**, **permanent dipole–permanent dipole interactions** and **hydrogen bonding**.

TOP TIP
Intramolecular forces refer to the bonds between **atoms** in a molecule, whereas intermolecular forces act between **molecules**.

London dispersion forces

London dispersion forces exist as a result of attractions between induced temporary dipoles in molecules. This is explained in Periodicity 1 (pages 18 and 19).

Permanent dipole–permanent dipole interactions

Permanent dipole–permanent dipole interactions occur between **polar molecules**. A molecule is only polar if it has a slightly positive (δ^+) side and a slightly negative side (δ^-). Polar molecules are said to have a permanent dipole.

TOP TIP
A molecule can contain polar bonds but the molecule need not necessarily be polar. Both the atoms in the molecule and its shape have to be considered.

Polar molecules

$\delta^+ H \!-\! Cl^{\delta-}$

$\delta+$ side of | $\delta-$ side of
molecule | molecule

(hydrogen chloride)

$\delta-$ side of molecule

$\delta-$O

H $\delta+$ H $\delta+$ $\delta+$ side of molecule

(water)

Non-polar molecules

$\delta-$O$=$C$^{\delta+}=$O$^{\delta-}$

(carbon dioxide)

$\overset{\delta-}{Cl}$

$\overset{}{C^{\delta+}}$

Cl   Cl
$\delta-$ $\delta-$

$\overset{}{\underset{\delta-}{Cl}}$

(tetrachloromethane)

The figures above clearly show that hydrogen chloride and water are polar molecules – they have a permanent dipole. Carbon dioxide and tetrachloromethane are non-polar molecules – they do not have dipoles. Both carbon dioxide and tetrachloromethane have a symmetrical shape, so the polarity of the bonds in a non-polar molecule cancel each other out and the molecules do not have a slightly negative (δ^-) side or slightly positive side (δ^+).

As one end of a polar molecule is slightly negative, it can form an attraction with the slightly positive end of a neighbouring molecule. This is the permanent dipole–permanent dipole interaction. Although these forces of attraction are stronger than London dispersion forces they are still relatively weak. The diagram on the right shows the permanent dipole–permanent dipole interactions between trichloromethane molecules.

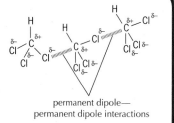

permanent dipole—
permanent dipole interactions

It isn't possible to tell by looking at two liquids to tell if they are polar or not, but a simple experiment can be carried out to tell them apart:

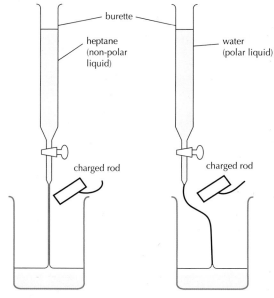

TOP TIP

Although London dispersion forces act between all molecules, permanent dipole–permanent dipole attractions are stronger, so generally have more of an effect on the properties of a substance.

Polar substances, such as water, are attracted to a charged plastic rod but non-polar substances, like heptane, are not.

Quick Test 11

1. Bromine (Br_2) and iodine monochloride (ICl) have the same number of electrons yet iodine monochloride has a higher melting point.

   Explain this observation by reference to the type of intermolecular forces between the molecules.

2. (a) The shape of a PCl_3 molecule is shown:

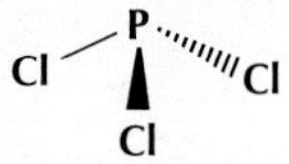

   (i) Describe the type of bonding between phosphorus and chlorine atoms in phosphorus trichloride (PCl_3).

   (ii) Is the molecule polar or non-polar?

   (iii) Justify your answer.

   (b) State what type of intermolecular force of attraction would predominate between PCl_3 molecules.

Structure and bonding 3

Hydrogen bonding

When hydrogen is bonded to the highly electronegative elements fluorine, oxygen and nitrogen, it results in the electrons in the bond being more strongly attracted away from the hydrogen atom. This would not be the case if other non-metal atoms were attached to the hydrogen. The hydrogen atom is so small that the positive charge on the atom is unusually high and the bond is highly polar. This results in a very strong permanent dipole–permanent dipole interaction between molecules called a **hydrogen bond**.

Hydrogen bonds are the **strongest** of the van der Waals forces, and they cause some unusual and unexpected properties in compounds in which the hydrogen bonds are present.

Hydrogen fluoride (HF), water (H_2O) and ammonia (NH_3) are compounds with hydrogen bonds between their molecules. A diagrammatic representation of hydrogen bonding in water is shown below:

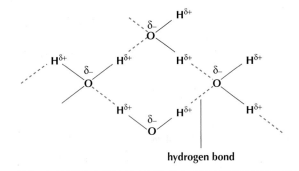

hydrogen bond

TOP TIP

Hydrogen bonding is a special type of permanent dipole–permanent dipole intermolecular attraction. It is the strongest van der Waals force and only occurs between molecules which have hydrogen bonded to oxygen, nitrogen or fluorine.

The effects of hydrogen bonding

Boiling points and melting points

These graphs show the boiling points and melting points of the hydrides of the elements in groups 4 to 7:

- Groups 5, 6 and 7: NH_3, H_2O and HF have higher boiling points than other hydrides in their periodic group. This is because they have **hydrogen bonding** between their molecules. The molecules of the other hydrides in the groups are held together by permanent dipole– permanent dipole interactions and London dispersion forces, which are not as strong as hydrogen bonds. Therefore, less energy is required to separate them.

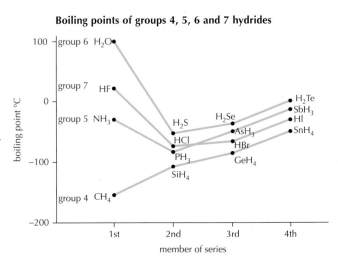

Boiling points of groups 4, 5, 6 and 7 hydrides

- Group 4 hydrides: the molecules are **non-polar** and are held together by weak London dispersion forces, which do not need much energy to separate them. As the molecules get bigger, they have more electrons, so the London dispersion forces increase. This in turn increases the boiling point of the compounds.

- The melting points show a similar trend, and can be explained in the same way.

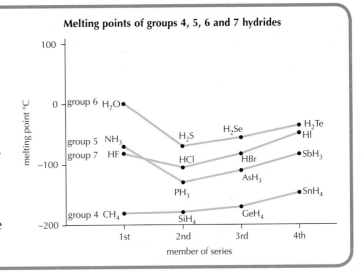

Melting points of groups 4, 5, 6 and 7 hydrides

The properties of water

Hydrogen bonding also gives water some unusual properties.

Density

Water is unusual in that when it is cooled to 4°C it begins to **expand**. At its freezing point it is less dense than liquid water – this explains why ice cubes and icebergs float on water. It also explains why water pipes can burst in winter. When the water freezes and expands, the force causes to pipes to crack. This expansion of water is due to the ordering of the molecules into an open structure, with an increased number of hydrogen bonds. This is less dense than the arrangement of the molecules in liquid water.

Viscosity

The type of bonding present between molecules affects the viscosity (thickness) of a liquid. This can be demonstrated by dropping steel balls through a variety of liquids.

- Diethyl ether is the least viscous of the liquids because it only has weak London dispersion forces holding molecules together.

- Ethanol molecules have hydrogen bonding between molecules, so are more viscous than diethyl ether, but less viscous than water. Water has more highly polar –OH groups

steel ball

substance:	water	ethanol	diethyl ether
structure:	$\overset{\delta-}{O}$ $\underset{\delta+}{H}$ $\underset{\delta+}{H}$	$\overset{\delta-}{O}$ $\underset{\delta+}{H}$ C_2H_5	O C_2H_5 C_2H_5
formula mass:	18	46	74

than ethanol, allowing more strong hydrogen bonds to form between water molecules than between ethanol molecules.

Solubility

Water's ability to dissolve a wide range of substances is due to the high polarity of the bonds within it. Water is particularly efficient at dissolving **polar** and **ionic** substances.

- Hydrogen chloride (HCl) exists as polar molecules. The polar covalent bond between the hydrogen and the chlorine breaks, and the resulting ions are attracted to the water molecules. The (aq) symbol is used to indicate that the ions are hydrated in solution:

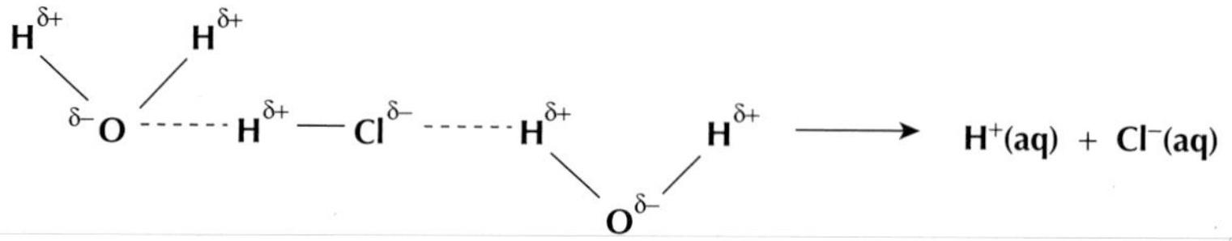

$$H^{\delta+}(aq) + Cl^{-}(aq)$$

- Sodium chloride exists as an ionic lattice. The water molecules interact with the ions in the lattice, breaking the electrostatic attraction between the ions. The ions then go into solution. A force of attraction is created between the ions and the polar water molecules – the ions are hydrated with water molecules.

- **Non-polar** substances like fats and oils cannot form strong intermolecular forces of attraction with water, and therefore are not dispersed throughout the water like soluble polar covalent and ionic substances. Non-polar solvents like tetrachloromethane can dissolve non-polar substances by forming London dispersion forces between the molecules.

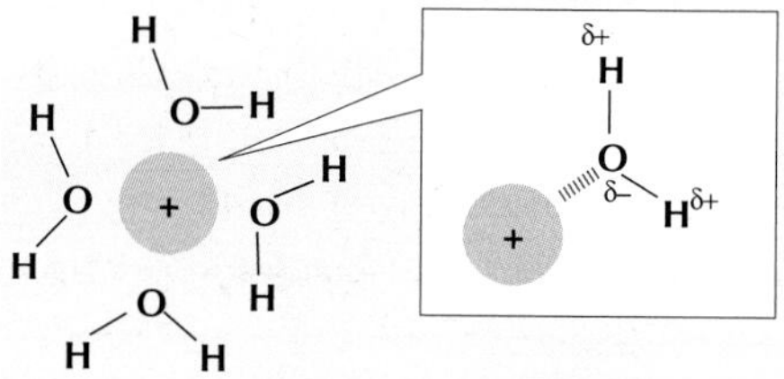

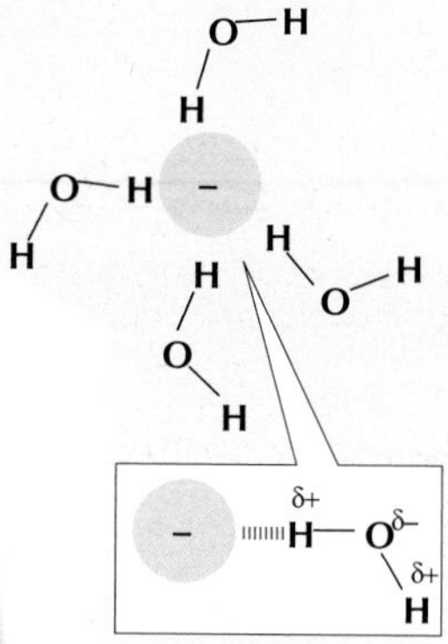

TOP TIP

Polar covalent and ionic substances generally tend to be soluble in polar solvents, like water. Non-polar substances generally tend to be soluble in non-polar solvents.

Miscibility

Miscibility is the ability of liquids to mix in all proportions, forming a solution. Water and ethanol, for example, are miscible and this forms the basis for alcoholic drinks.

The water and ethanol molecules are both polar and they can hydrogen-bond with each other, which allows them to mix completely.

Oil and water are two liquids which are well known to be **immiscible** – they form two separate layers when mixed. Oil is non-polar, and so does not form forces of attraction with the polar water molecules. This means they separate when shaken together.

Quick Test 12

1. (a) Identify which of these compounds have hydrogen bonding between molecules:

    (i) HBr;   (ii) NH_3;   (iii) HF;   (iv) CH_3OH

    (b) Justify your choices.

2. (a) This table shows the boiling points of the group 6 hydrides, and the number of electrons in a molecule of each compound.

Group 6 hydride	H_2O	H_2S	H_2Se	H_2Te
Number of electrons per molecule	10	18	36	54
Boiling point (°C)	100	–60	–41	–2

    (i) Based on the number of electrons in a molecule of each of the hydrides, what might the boiling point of H_2O have been predicted to be?

    (ii) Explain why the boiling point of H_2O does not follow the expected trend shown by the other group 6 hydrides.

    (b) When water freezes in pipes during the winter, it can cause them to burst. Explain what has happened to the water to cause this.

3. (a) Explain why potassium chloride dissolves easily in water but candle wax, which is a mixture of large hydrocarbon molecules, does not dissolve in water.

    (b) Heptane (C_7H_{16}) is a solvent that dissolves candle wax; it is a mixture of large hydrocarbon molecules. Explain why heptane can dissolve candle wax but water cannot.

Learning checklist

Controlling the rate of reaction

In this section you have learned:

- For reactants to form products they must first come in contact with each other (collide).
- The higher the concentration, the more collisions, and so the faster the reaction.
- The smaller the particle size, the bigger the surface area so the higher the chance of collision.
- Before collisions can be successful and lead to products, a minimum amount of energy, known as the activation energy (E_a), is required.
- Increasing the temperature increases the number of particles with energy equal to or greater than E_a. This results in more successful collisions and an increase in rate.
- Energy distribution diagrams can be drawn to show how increasing the temperature results in more particles with energy equal to or greater than E_a.
- Relative rate of reaction is inversely proportional to time, i.e. rate = 1/t.
- Rate of reaction is directly proportional to the concentration of the reactants.
- Rate of reaction is not directly proportional to temperature – a small increase in temperature results in a large increase in rate.
- Collision geometry has an effect on reaction rate.
- Reaction profiles can be drawn to show how the potential energy changes as a reaction progresses.
- Enthalpy is the name given to stored energy in chemicals and is given the symbol H.
- ΔH is the change in enthalpy for a reaction, and is measured in kJ or kJ mol^{-1} when one mole of substance is reacted or produced.
- ΔH and E_a can be identified in a reaction profile for exothermic and endothermic reactions.
- ΔH values are negative for exothermic reactions and positive for endothermic reactions.
- E_a is shown as an energy barrier in a reaction profile.
- An activated complex is formed at the top of the potential energy barrier.
- Catalysts provide an alternative pathway for reactants to form products – each step has a lower E_a than the uncatalysed reaction.

Periodicity

In this section you have learned:

- The first 20 elements in the periodic table are categorised according to their bonding and structure:
  - metallic (Li, Be, Na, Mg, Al, K, Ca)
  - monatomic (He, Ne, Ar)
  - covalent molecular (H_2, N_2, O_2, F_2, Cl_2, P_4, S_8 and fullerenes)
  - covalent network [B, C (diamond and graphite), Si]

Non-metals
Monatomic
- The noble gases are monatomic elements with weak London dispersion forces between the atoms.
- London dispersion forces are weak forces of attraction between atoms and molecules when temporary dipoles are formed within atoms or molecules.

Covalent molecular
- Hydrogen, nitrogen, oxygen and the halogens exist as individual diatomic molecules, with London dispersion forces between the molecules (intermolecular), and strong covalent bonds between the atoms in a molecule (intramolecular).
- Sulfur and phosphorus are small individual molecules with strong London dispersion forces between molecules, so are solids at room temperature.
- Carbon can exist as large individual covalent molecules (fullerenes), which have molecules with 60 carbon atoms or more.
- Fullerene molecules can be ball or tube-shaped, and have high melting points because of the large London dispersion forces between molecules.

Covalent network
- Carbon can exist as large covalent – for example, diamond and graphite.
- The carbon atoms in diamond are bonded to four other carbons in a large, strong three-dimensioal network. This makes it very hard.
- The network structure of diamond is results in diamond having a very high melting point.
- Diamond is a non-conductor of electricity because it has no delocalised electrons.
- Graphite forms a network in which each carbon is only bonded to three other carbons, which results in graphite having delocalised electrons, so therefore conducts electricity.
- The carbon atoms in graphite form hexagonal plates which are held together by London dispersion forces.
- The layers in graphite can be easily separated, which results in graphite not being as hard as diamond.

- Boron and silicon exist as covalent networks, and have properties associated with covalent networks, e.g., very high melting and boiling points.

Trends in the periodic table

- Patterns in changes of covalent radii, ionisation energy and electronegativity exist when going down a group, or across a period in the periodic table.
- The covalent radius is half the distance between the nuclei of two atoms, joined by a covalent bond.
- The covalent radius increases down a group as the number of occupied electron shells increases, and shields the outer electrons from the pull of the nucleus.
- The covalent radius decreases going across a group as there is no increased shielding effect. The electrons are added to the same electron shell.
- Ionisation energy is the energy required to remove one mole of electrons from one mole of gaseous atoms.
- First ionisation energy decreases down a group as the outer electrons get further from the nucleus and are shielded from the pull of the nucleus by the shells of the inner electrons.
- First ionisation energy generally increases across a period as the outer electrons are held more tightly by the nucleus as electrons are added to the same energy level and the charge on the nucleus increases.
- Electronegativity is a measure of the attraction an atom has for the electrons of the bond.
- Electronegativity decreases down a group as atoms get bigger, and the attraction of the nucleus for bonding electrons decreases.
- Electronegativity increases across a period as the atoms get smaller, and the attraction of the nucleus for bonding electrons increases.

Structure and bonding

In this section you have learned:

- In pure covalent bonding, the shared pair of electrons is shared equally by the atoms in the molecule.
- In polar covalent bonding there is unequal sharing of the bonded electrons, which results in one atom having a slightly negative charge (δ^-), and the other having a slightly positive charge (δ^+). This creates a permanent dipole.
- Electronegativity is a numerical measurement on a scale of 0–4 (low to high) used to assess an element's ability to attract bonding electrons.
- The difference in electronegativity between the atoms in a compound gives an indication of the type of bonding present.
- The type of bonding changes gradually as the difference in electronegativity between atoms increases – this is known as the bonding continuum.

- The bonding continuum has ionic bonding at one end, and pure covalent bonding at the other. There is polar covalent bonding between the two extremes.
- Intermolecular forces act between molecules, and are known as van der Waals forces.
- London dispersion forces, permanent dipole–permanent dipole interactions, and hydrogen bonding are the three types of van der Waals forces.
- London dispersion forces are weak attractions between temporary dipoles in molecules, and are significant between non-polar molecules.
- Permanent dipole–permanent dipole interactions are attractions between polar molecules which have permanent dipoles.
- A molecule will only be polar if one side of the molecule is slightly negative (δ^-) and the other slightly positive (δ^+).
- Hydrogen bonds form between highly polar molecules which have hydrogen bonded to the greatly electronegative elements fluorine, oxygen or nitrogen.
- Hydrogen bonding gives compounds unusually high melting and boiling points. This is because they are the strongest van der Waals force, and so need more energy to separate the molecules, e.g. water.
- When water is cooled to 4°C or below, it expands and decreases in density. This is due to the water molecules forming a more open structure, with increased hydrogen bonding.
- Compounds with molecules which have hydrogen bonding between them are more viscous than compounds with molecules which have other intermolecular forces.
- The polar nature of water makes it a good solvent for ionic compounds and polar covalent molecular compounds.
- Non-polar solvents can dissolve non-polar substances.
- The molecules of polar covalent liquids which are miscible with water form hydrogen bonds with water molecules.

Esters, fats and oils 1

Making and using esters

Esters are formed by the **condensation** reaction between a carboxylic acid and an alcohol. Concentrated sulfuric acid acts as a catalyst.

Many fruits get their distinctive flavour from esters. Artificial esters are widely used in the confectionary industry as flavouring for sweets. Some are used in perfumes because of their strong, pleasant smell. Others, like ethyl ethanoate, are used as solvents for products such as of nail varnish.

Naming esters

The first part of the name is derived from the alcohol and ends in **-yl**; the second part comes from the acid and ends in **-oate**.

TOP TIP

Although the first part of the name comes from the alcohol, the acid part of the ester is usually drawn on the left. Notice the way the structure of the alcohol is drawn to show how the hydroxyl group reacts with the carboxyl group.

Example

methanoic acid + ethanol ⇌ eth**yl** methan**oate** + water

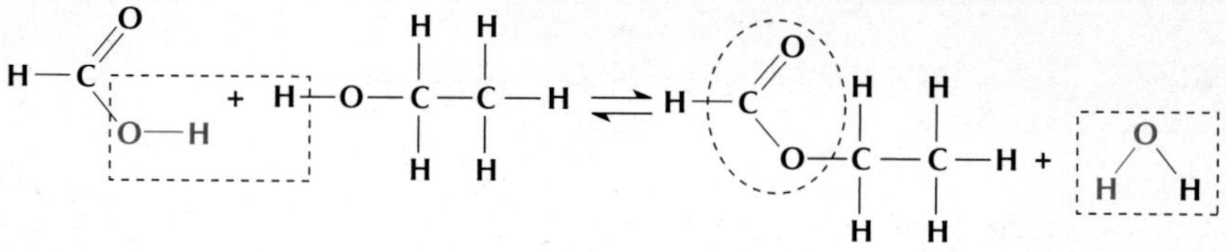

Note: (1) 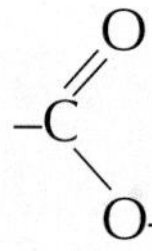 the ester link structure is known as the **ester link**.

(2) The ester link is formed when the carboxyl group of the acid reacts with the hydroxyl group of the alcohol.

(3) **Water** is formed during the reaction.

(4) The ⇌ symbol indicates the reaction is reversible.

TOP TIP

Given the names of the parent carboxylic acid and alcohol or their structural formulae, you should be able to **name** the ester formed when they react.

TOP TIP

Given the names of the parent carboxylic acid and alcohol or the names of esters, you should be able to **draw** the structural formulae for esters.

Breaking esters

The formation of esters is reversible, so esters can be broken down into their parent acid and alcohol by reacting with water – this is known as **hydrolysis**. This requires the ester to be heated with an acid or alkali to speed up the process.

When a dilute acid is used, the parent carboxylic acid and parent alcohol are obtained. When ethyl propanoate is hydrolysed using dilute acid, propanoic acid and ethanol are obtained.

> **TOP TIP**
>
> Given the name or the structural formula of an ester, the hydrolysis products can be named and their structural formulae drawn.

$$
\underset{\text{ethyl propanoate}}{
\begin{array}{c}
\text{H} \ \ \text{H} \ \ \text{O} \\
| \ \ \ | \ \ \ \| \\
\text{H}-\text{C}-\text{C}-\text{C}-\text{O}-\text{C}-\text{C}-\text{H} \\
| \ \ \ | \ \ \ \ \ \ \ \ | \ \ \ | \\
\text{H} \ \ \text{H} \ \ \ \ \ \ \text{H} \ \ \text{H}
\end{array}} +
\underset{\text{water}}{
\begin{array}{c}
\text{O} \\
/ \ \ \backslash \\
\text{H} \ \ \ \ \text{H}
\end{array}} \rightleftharpoons
\underset{\text{propanoic acid}}{
\begin{array}{c}
\text{H} \ \ \text{H} \ \ \text{O} \\
| \ \ \ | \ \ \ \| \\
\text{H}-\text{C}-\text{C}-\text{C}-\text{O}-\text{H} \\
| \ \ \ | \\
\text{H} \ \ \text{H}
\end{array}} +
\underset{\text{ethanol}}{
\begin{array}{c}
\text{H} \ \ \text{H} \\
| \ \ \ | \\
\text{H}-\text{O}-\text{C}-\text{C}-\text{H} \\
| \ \ \ | \\
\text{H} \ \ \text{H}
\end{array}}
$$

The molecule always splits at the ester link, as shown by the dotted line.

Quick Test 13

1. (a) Name the esters formed when the following carboxylic acids and alcohols react:

    (i) methanoic acid and propan-1-ol.

    (ii)
$$
\begin{array}{c}
\text{H} \ \ \text{H} \ \ \text{H} \ \ \text{O} \\
| \ \ \ | \ \ \ | \ \ \ \| \\
\text{H}-\text{C}-\text{C}-\text{C}-\text{C}-\text{O}-\text{H} \ + \ \text{H}-\text{O}-\text{C}-\text{C}-\text{H} \\
| \ \ \ | \ \ \ | \ \ \ \ \ \ \ \ \ \ \ \ \ \ \ \ \ \ \ | \ \ \ | \\
\text{H} \ \ \text{H} \ \ \text{H} \ \ \ \ \ \ \ \ \ \ \ \ \ \ \ \text{H} \ \ \text{H}
\end{array}
$$

    (b) Draw the structural formulae for the esters formed.

2. Name the products formed when the following esters are hydrolysed under acid conditions, and draw their structural formulae.

    (a) ethyl ethanoate

    (b)

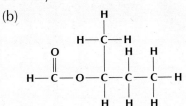

Esters, fats and oils 2

TOP TIP

Hydrolysis of a fat or oil will always give one mole of glycerol molecules to three moles of fatty acid molecules. The three fatty acids in a fat or oil molecule may be the same or different.

Fats and oils

Fats and oils are naturally occurring esters which come from animals (e.g. butter and fish oil) and plants (e.g. olive oil). They are a concentrated source of energy, and are essential for the storage and transport of fat-soluble vitamins in the body. Fats and oils are formed by a condensation reaction between long-chain carboxylic acids, often called **fatty acids**, and **glycerol** (propane-1,2,3-triol).

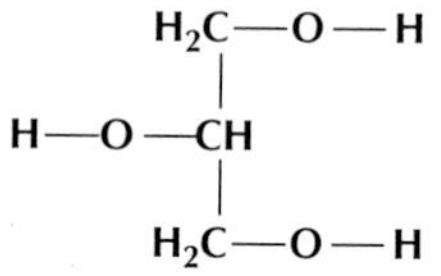

glycerol (propane-1,2,3-triol)

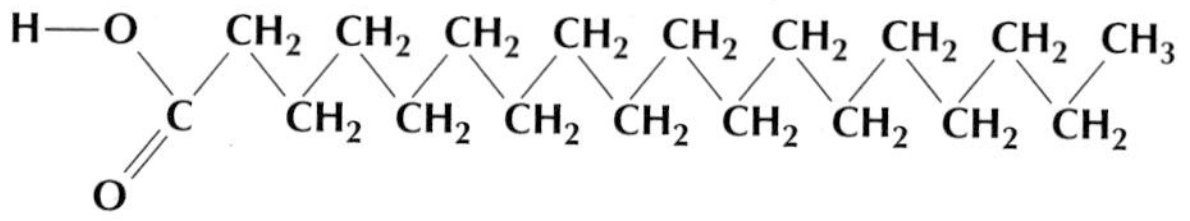

stearic acid (octadecanoic acid), a fatty acid

The hydrocarbon chain can be represented as a serrated line, or by the letter R, in order to simplify the structure of a fatty acid.

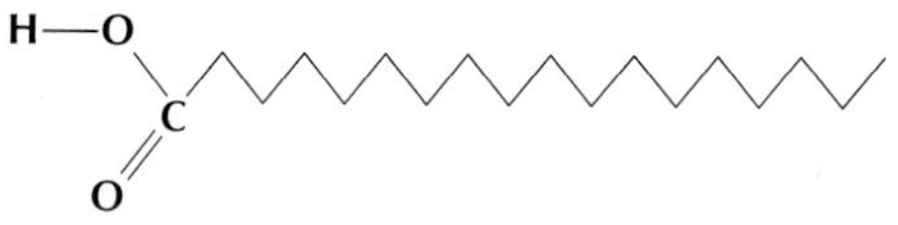

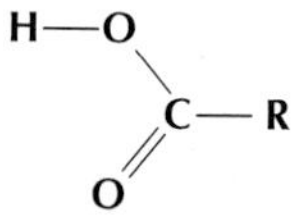

The hydrocarbon tail can be saturated or partially unsaturated, giving rise to a variety of fatty acids. The different hydrocarbon tails can be represented as R^1, R^2, R^3 and so forth.

Glycerol has three hydroxyl groups in each molecule and so reacts with three fatty acid molecules. This results in each fatty acid molecule forming an ester link with the glycerol molecule. This is why fats and oils are sometimes called triglycerides. One of the ester links in a triglyceride molecule is shown within the dotted area in the diagram on the right.

If the fatty acid chains are mostly **saturated**, the substance will likely be solid at room temperature and be called a **fat**. The greater the proportion of **unsaturated** fatty acid chains, the more likely the triglyceride is to be liquid at room temperature. In this case, it is called an **oil**.

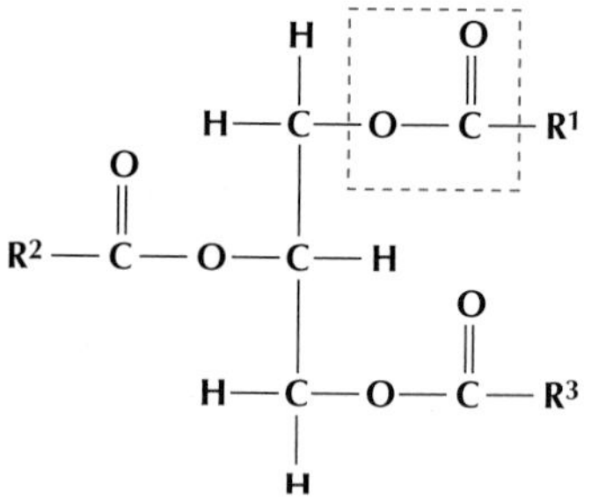

A simplified triglyceride molecule, highlighting one of the ester links in the molecule.

The structure of fats and oils

In **fats**, the shape of the molecules allows them to pack closer together with stronger van der Waals forces between them than is the case with oils. This means fats need more energy to separate the molecules, and are therefore solids at room temperature.

In **oils**, the presence of double bonds on the fatty acid parts of the molecules causes the hydrocarbon chains of the fatty acids to be more kinked. This means the molecules are unable to pack as closely together. This indicates that the van der Waals forces are weaker, and so it takes less energy to separate them. Triglycerides with double bonds have a lower melting point, and tend to be liquids (oils).

This is shown in the simplified diagrams below.

A simplified 'tuning fork' representation of fat molecules showing the van der Waals attractions between molecules.

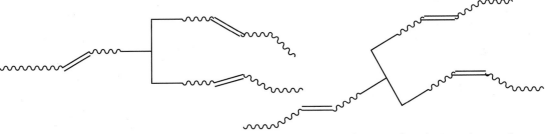

The double bonds in oil molecules 'kink' the tail of the molecule, which means they don't pack as easily as fat molecules.

The fats in butter are mostly saturated while oils are mostly unsaturated.

TOP TIP

Fat and oil molecules have very complicated structures. Simplified structures are used to help explain how they are formed and to explain their properties. Fats and oils are esters – look out for the ester link in their structures.

Quick Test 14

1. Explain why a fat molecule has three ester links.
2. Suggest a chemical test that you could use to show that oil molecules are unsaturated.
3. (a) Explain why fats are solid at room temperature while oils are liquid.
    (b) Suggest why some oils go solid if placed in the fridge.

Proteins and enzymes

Proteins

Proteins are needed in our diet for growth and repair of body tissue, for example hair, nails and muscle. Proteins are also responsible for the maintenance and regulation of life processes.

Amino acids

Proteins are natural condensation polymers made up of **amino acids**. Each amino acid has two functional groups, an **amino group**, $-NH_2$ and a **carboxyl group, –COOH**.

The structure of an amino acid can be represented as:

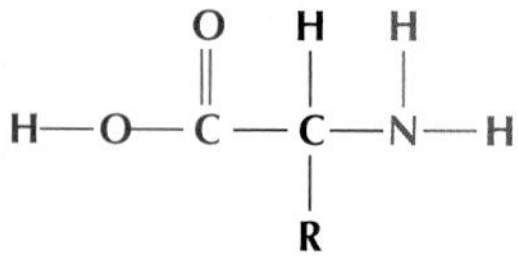

TOP TIP

If you are asked to show how amino acids join, disregard the R groups and focus on where the carboxylic acid and amino acid groups are.

R varies and can be, for example, H, CH_3 etc. and can also contain S, N and O.

There are 20 amino acids needed by the body to make proteins. These amino acids combine in varying amounts and in different sequences. The body cannot make all the amino acids needed, and so relies on food for these **essential amino acids**.

When two amino acids join, a water molecule is also produced. The diagram below shows three amino acids joining to form part of a protein structure:

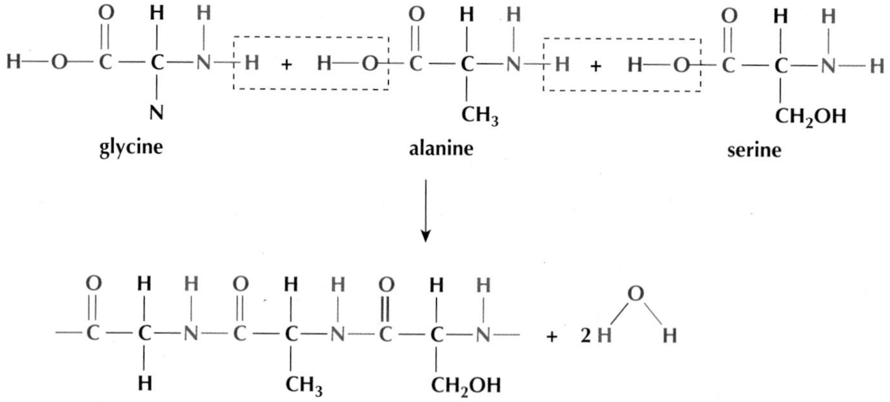

glycine   alanine   serine

TOP TIP

A protein can be identified by the presence of the peptide link in the structure:

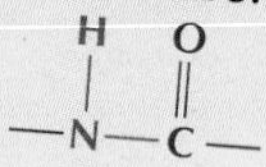

part of a protein molecule

The new link formed between the amino acids is known as the **peptide** (or amide) **link**.

Enzymes

Enzymes are responsible for the many chemical reactions which take place in our bodies. Most enzymes are proteins. Enzymes are described as **biological catalysts** because they

speed up biochemical reactions, as well as the digestion of dietary protein. Enzymes catalyse the breakdown of chemicals found in foods.

Enzyme	Breaks down
amylase	carbohydrate
pepsin and trypsins	protein
lipases	fats

When dietary protein is **hydrolysed** in the body amino acids are formed. The hydrolysis takes place at the peptide link.

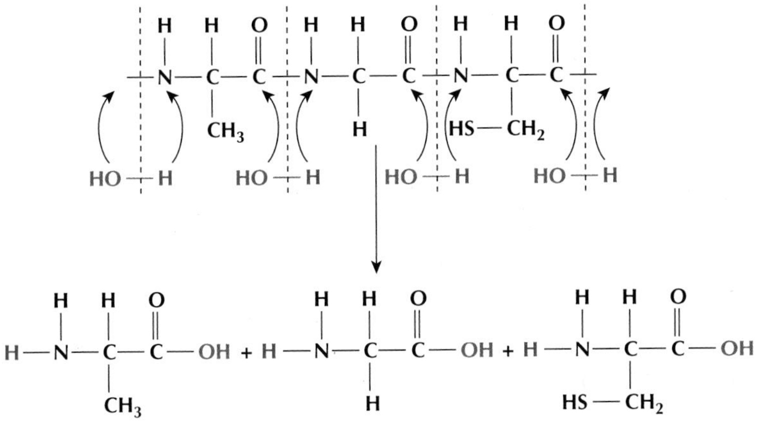

TOP TIP

Given part of a protein structure, you should be able to draw the amino acids obtained by hydrolysis of the protein. Hydrolysis always takes place at the peptide link – H always adds to the amino acid part and OH to the carboxylic acid part.

Quick Test 15

1. (a) (i) Draw part of the protein structure formed when the three amino acids shown below join.

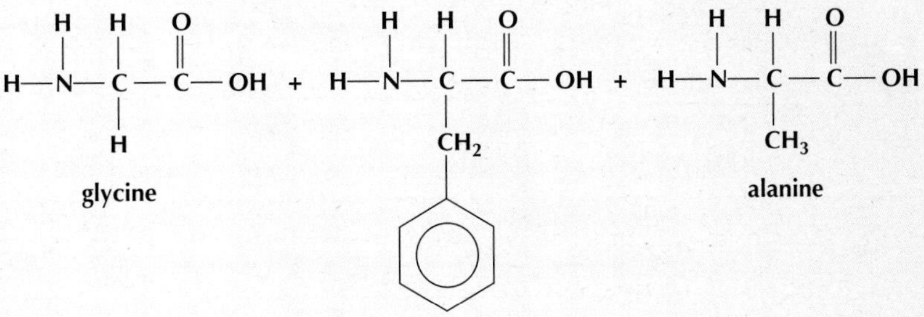

glycine   phenylalanine   alanine

(ii) Circle the peptide links in the structure.

2. Draw the structural formulae for the amino acids formed when the protein structure shown below is hydrolysed.

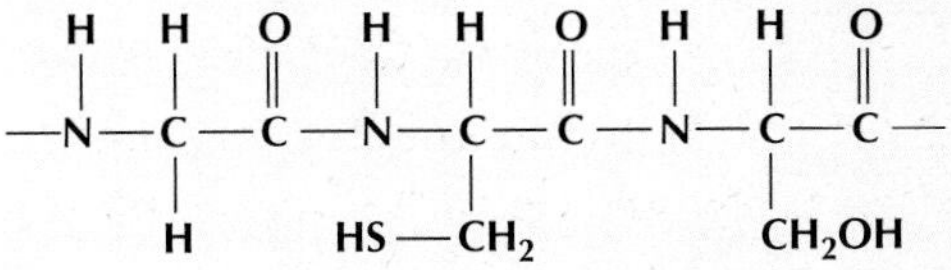

The chemistry of cooking 1

Denaturing protein

Proteins are long chain molecules that can be twisted to form spirals and folded into sheets. The chains are held in these forms by hydrogen bonding within the molecule. When heated, these forces of attraction are broken and the protein is **denatured** – it loses its shape.

The white of an egg (albumen) is a globular protein – the chains are clumped together in small balls. When heated, the strands unwind as hydrogen bonds are broken. New bonds form between protein strands, resulting in the familiar white colour and texture of cooked egg.

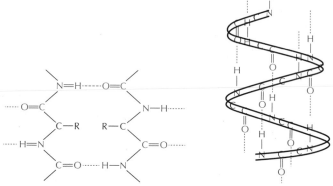

Hydrogen bonding within a protein molecule holds it in a particular shape.

Aldehydes and ketones

The flavour in foods is the result of a complex mixture of chemicals, including **aldehydes** and **ketones**, which both contain the **carbonyl group**:

Aldehydes always have the carbonyl functional group at the end of the structure and at least one hydrogen attached to the carbonyl group. Ketones have no hydrogens attached to the carbonyl group – there are always two carbons attached. Aldehydes have the name ending –al. Ketones have the name ending –one. Both take the first part of their name from the corresponding alkane. See the table below for examples of an aldehyde and a ketone.

Name	Structural formula	Shortened structural formula	Molecular formula
propan**al**		CH_3CH_2CHO	C_3H_6O
propan**one**		CH_3COCH_3	C_3H_6O

Boiling points

Carbonyl compounds are polar. They contain a dipole along the carbon–oxygen double bond. This creates permanent dipole–permanent dipole attractions between molecules. These forces of attraction result in aldehydes and ketones having boiling points that are higher than their equivalent alkane (due to weaker London dispersion forces) but lower than their corresponding alcohol (due to stronger hydrogen bonding). Small aldehydes and ketones are very volatile, so even gently warming them causes them to emit odours often associated with cooking certain foods.

Solubility

Carbonyl compounds can form hydrogen bonds with water molecules, meaning that lower molecular mass aldehydes and ketones are soluble (miscible) in water.

TOP TIP

Although you will not be asked to name complex aldehydes and ketones, you may be given a name or structural formula and asked to identify it as either an aldehyde or ketone.

Some foods, like asparagus, have water-soluble flavour molecules so they lose flavour if boiled. Others, like broccoli, have flavour molecules which are more soluble in oil so they don't lose flavour by cooking them in water.

Quick Test 16

1. The carvone molecule shown is what gives spearmint its smell and flavour. It is used in the manufacture of some spearmint gums.

carvone

   (a) Circle the carbonyl group in the structure.

   (b) State what type of molecule carvone is, and justify your answer.

   (c) What does the use of carvone suggest about its volatility?

2. (a) Draw the structural formulae of the straight chain aldehyde and ketone with five carbon atoms and write their molecular formulae.

   (b) What term is used to describe molecules with the same molecular formula, but different structures?

The chemistry of cooking 2

Systematically naming branched-chain aldehydes and drawling structural formulae

Using the systematic naming rules, given the structural formula:

1. name the longest hydrocarbon chain;
2. name the branches (side chains), e.g. CH_3 -methyl; C_2H_5 -ethyl; etc.;
3. Indicate the position of the branches on the main chain, numbering from the carbonyl group as carbon 1;

   prefixes are used if there is more than one side chain of the same type (e.g. di- is used if there are two of the same type, tri- if there are three, and so forth).

Example:

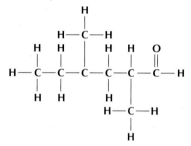

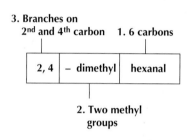

3. Branches on 2nd and 4th carbon   1. 6 carbons

| 2, 4 | – dimethyl | hexanal |

2. Two methyl groups

Given the systematic name, you can draw the structural formulae for branched-chain aldehydes:

1. identify the number of carbon atoms in the longest straight-chain;
2. identify the branches and the numbers of the carbon atoms to which they are attached.

Naming ketones from structural formulae

Straight-chain ketones are named in a similar way to alcohols, except their suffix is -one.

Example:

pentan-2-one   pentan-3-one

Systematically naming branched-chain ketones and drawing structural formula

Systematic naming rules, given the structural formula:

1. name the longest carbon chain containing the carbonyl group;
2. number the chain from the end that the carbonyl group is nearer to, and identify the position of the carbonyl group;
3. name the branches (side chains), e.g. CH_3 -methyl; C_2H_5 -ethyl; etc. **(Prefixes are used if there is more than one side chain of the same type (e.g. di- is used if there are two of the same type, tri- if there are three, etc.);**
4. indicate the position of the branches on the main chain.

TOP TIP

The branches are arranged in alphabetical order within the name.

Example:

4. Branches on 3rd and 4th carbon

1 and 2: Six carbons with carbonyl on carbon

| 3 – | ethyl | – 4 – | methyl | hexan – 2 – one |

3. Two branches

Given the systematic name, you can draw the structural formulae for branched-chain aldehydes.

1. identify the number of carbon atoms in the longest straight-chain containing the carbonyl group;
2. identify the position of the carbonyl group;
3. identify the branches and the numbers of the carbon atoms to which they are attached.

Quick Test 17

1. (a) Draw structural formulae for 3-methylpentanal and 4-methylpentan-2-one.

   (b) Write the molecular formula for both structures.

   (c) A student stated that the two compounds in (a) couldn't be isomers because they belong to different homologous series. Comment on this statement.

2. Write the systematic names for structures A and B.

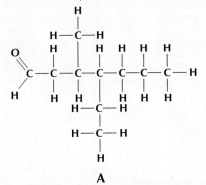

A

B

The chemistry of cooking 3

Oxidation of aldehydes

Although aldehydes and ketones both contain the carbonyl functional group, aldehydes can be **oxidised** whereas ketones cannot.

When an aldehyde is oxidised a carboxylic acid is formed. For example, ethanal would be oxidised to ethanoic acid.

The ion-electron equation for this reaction is:

$$CH_3CHO(\ell) \ + \ H_2O(\ell) \ \rightarrow \ CH_3COOH(\ell) \ + \ 2H^+(aq) \ + \ 2e^-$$

As electrons appear on the right-hand side of the equation, i.e. are lost from the reactants, the equation shows that the reaction is an oxidation.

Distinguishing aldehydes from ketones

The fact that aldehydes can be oxidised but ketones can't is useful for telling them apart. Three oxidising agents can be used to distinguish aldehydes and ketones: acidified potassium dichromate, Fehling's solution and Tollens' reagent. A few drops of the compound to be tested are added to the chosen reagent in a test tube, and the test tube placed in a hot water bath. Each of the reagents gives a distinctive result with aldehydes only.

Acidified potassium dichromate

Orange dichromate ions are **reduced** to green chromium(III) ions.

$$Cr_2O_7^{2-}(aq) \ + \ 14H^+(aq) \ + \ 6e^- \ \rightarrow \ 2Cr^{3+}(aq) \ + \ 7H_2O(\ell)$$

orange   green

Fehling's solution

Fehling's solution is an alkaline solution containing blue copper(II) ions. The blue copper(II) ions are **reduced**, giving a precipitate of red/brown copper(I) oxide. The simplified equation is:

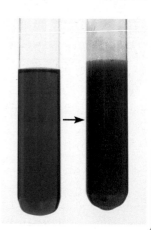

$$2Cu^{2+}(aq) \ + \ 2OH^-(aq) \ + \ 2e^- \ \rightarrow \ Cu_2O(s) \ + \ H_2O(\ell) \ + \ 2e^-$$

blue   red/brown

Tollens' reagent

Tollens' reagent is ammoniacal silver nitrate, and contains colourless silver(I) ions. Aldehydes **reduce** the silver ions to the metal and a very thin layer of silver metal is deposited on the test tube. For this reason, the test is often referred to as the 'silver mirror test'.

$$Ag^+(aq) \quad + \quad e^- \quad \rightarrow \quad Ag(s)$$
$$\text{colourless} \qquad\qquad\qquad \text{silver mirror}$$

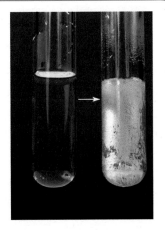

Quick Test 18

1. (a) Compounds A, B and C were known to be either aldehydes or ketones. The table shows the results of chemical tests carried out on the compounds. Fill in the missing answers (i)–(v).

Compound	Reagent	Test result	Aldehyde/Ketone?
A	(i)	orange → (ii)	aldehyde
B	Tollens'	no change	(iii)
C	(iv)	blue → red/brown	(v)

(b) Compounds B and C were found to have the molecular formula C_4H_8O. Draw the structural formulae for A and B, naming the compounds.

(c) Compound A is oxidised during the reaction. Name the type of compound formed when A is oxidised.

Oxidation of food 1

Chemicals in foods

In addition to aldehydes and ketones, foods can contain alcohols and carboxylic acids. These, and other food chemicals such as fats and oils, can be oxidised by oxygen in the air. Oxidation of food chemicals spoils the food, and so manufacturers have developed ways of preventing oxidation.

Systematically naming branched-chain alcohols

Alcohols have the hydroxyl functional group –O–H. Straight-chain alcohols take their name from the corresponding alkane. The -e is dropped and replaced by –ol.

Branched alcohols are named in the same way as ketones:

1. name the longest carbon chain containing the hydroxyl group;

2. number the chain from the end that the hydroxyl group is nearer to and identify the position of the hydroxyl group;

3. name the branches (side chains), e.g. CH_3 –methyl; C_2H_5 –ethyl, etc.;

    use prefixes if there is more than one side chain of the same type (e.g. di- is used if there are two of the same type, tri- if there are three, etc.);

4. indicate the position of the branches on the main chain;

    Example:

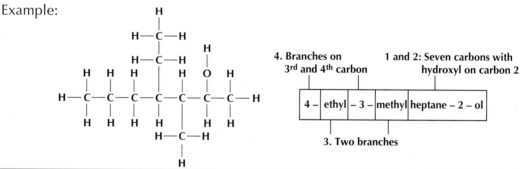

Drawing the structure of branched-chain alcohols from systematic names

1. identify the number of carbon atoms in the longest straight chain containing the hydroxyl group;

2. identify the position of the hydroxyl group;

3. identify the branches and the numbers of the carbon atoms to which they are attached.

Example:

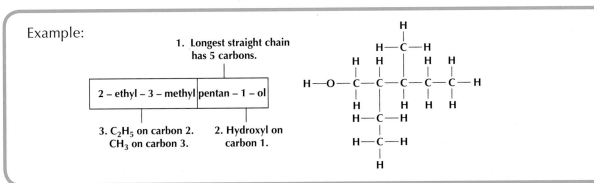

1. Longest straight chain has 5 carbons.

2 – ethyl – 3 – methyl pentan – 1 – ol

3. C_2H_5 on carbon 2. CH_3 on carbon 3.

2. Hydroxyl on carbon 1.

Boiling points of alcohols

The bar chart shows the boiling points of alcohols with up to four carbon atoms. They are:

CH_3OH
methanol

CH_3CH_2OH
ethanol

$CH_3CH_2CH_2OH$
propan-1-ol

$CH_3CH_2CH_2CH_2OH$
butan-1-ol

They are compared with the equivalent alkanes (methane to butane) with the same number of carbon atoms.

Notice that:

- The boiling point of an alcohol is always much higher than that of the alkane with the same number of carbon atoms.

- The boiling points of the alcohols increase as the number of carbon atoms increases.

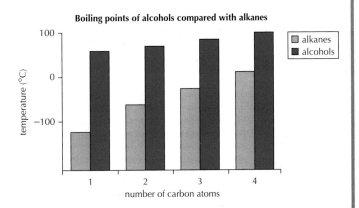

Boiling points of alcohols compared with alkanes

□ alkanes
■ alcohols

temperature (°C)

number of carbon atoms

The patterns in the boiling point reflect the patterns in intermolecular attractions. Alcohol molecules have the polar –O–H group, which results in hydrogen bonding between the molecules. Hydrogen bonds are stronger than the London dispersion forces which exist between the alkane molecules. This means it requires more energy to separate alcohol molecules from each other, and thus their boiling points are higher.

Quick Test 19

1. Write the systematic name for the structure shown:

2. Draw the structural formula for 4-methylhexan-3-ol.

3. Ethanol (bp: 78°C) and fluoroethane (bp: –37°C) have the same number of electrons in a molecule. Explain then why their boiling points are so different.

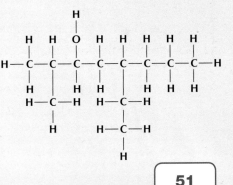

Oxidation of food 2

Classification of alcohols

Alcohols can be classified as primary, secondary and tertiary.

In **primary alcohols**, the carbon atom with the hydroxyl group is attached to **one** other carbon atom, e.g. propan-1-ol:

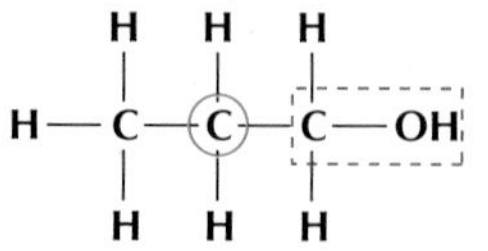

In **secondary alcohols**, the carbon atom with the hydroxyl group is attached to **two** other carbon atoms, e.g. propan-2-ol:

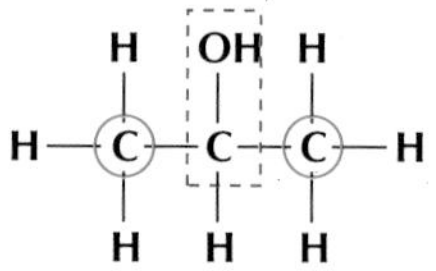

In **tertiary alcohols**, the carbon atom with the hydroxyl group is attached to **three** other carbon atoms, e.g. 2-methylpropan-2-ol:

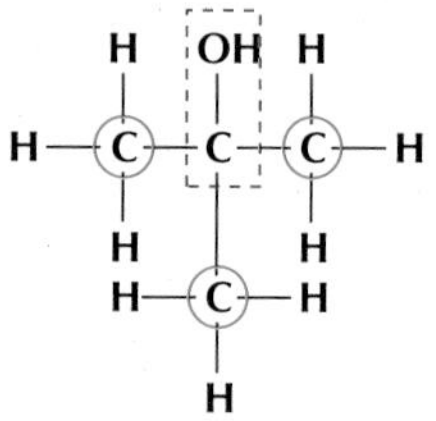

TOP TIP

It is important to be able to classify alcohols in order to predict whether they will react. Although you don't need to be able to name complex alcohols, you should be able to classify them from their structural formula.

Diols and triols

Alcohols with **two hydroxyl groups** are referred to as **diols**.

The simplest diol is ethane-1,2-diol:

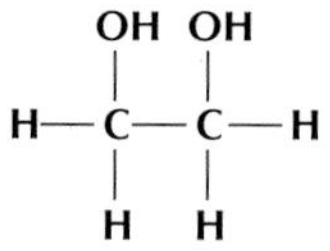

Alcohols with **three hydroxyl groups** are referred to as **triols**.

Glycerol is an example of a triol. Its systematic name is propane-1,2,3-triol. Glycerol is the alcohol part of fats and oils.

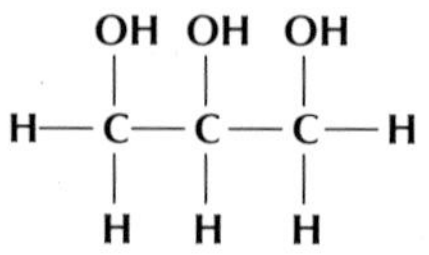

Diols and triols have high boiling points due to strong **hydrogen bonding** between molecules.

TOP TIP

The more hydroxyl groups an alcohol has, the stronger the hydrogen bonding. This means more energy is needed to separate the molecules, so the boiling point is higher.

Oxidation of alcohols

If a product molecule has a larger oxygen:hydrogen ratio than the reactant, then oxidation has taken place.

Primary alcohols can be oxidised to **aldehydes** then **carboxylic acids**.

Example:

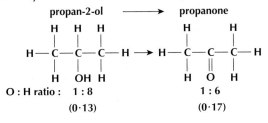

ethanol ⟶ ethanal ⟶ ethanoic acid

O : H ratio : 1 : 6   1 : 4   1 : 2
(0·17)   (0·25)   (0·50)

Secondary alcohols are oxidised to **ketones**.

propan-2-ol ⟶ propanone

O : H ratio : 1 : 8   1 : 6
(0·13)   (0·17)

TOP TIP

Confused about ratios? Think in fractions!
Remember 1:6 (1/6 = 0·17)
1:4 (1/4 = 0·25)
0·17 < 0·25
So, going from 1:6 to 1:4 represents an increase in the oxygen to hydrogen ratio.
Tertiary alcohols **cannot** be oxidised.

The oxidising agents are:

Hot copper(II) oxide – the copper ions are reduced: $Cu^{2+} + 2e^- \rightarrow Cu$

Acidified dichromate – the dichromate ion (orange) is reduced to the chromium(III) ion (green): $Cr_2O_7^{2-}(aq) + 14H^+(aq) + 6e^- \rightarrow 2Cr^{3+}(aq) + 7H_2O(\ell)$

Quick Test 20

1. Write the systematic name for the alcohol shown and state its classification:

2. Draw the structural formula for 2-methylbutan-2-ol and state its classification.

3. Describe what you would see if each of the alcohols in 1. and 2. were warmed with acidified dichromate solution, and explain your observations.

Oxidation of food 3

Systematically naming branched-chain carboxylic acids

Carboxylic acids have the carboxyl functional group –COOH at the end of the structure. Straight-chain carboxylic acids take their name from the corresponding alkane. The -e is dropped and replaced by –oic.

Branched alcohols are named in the same way as aldehydes:

1. name the longest hydrocarbon chain;
2. name the branches, e.g. CH_3 –methyl; C_2H_5. –ethyl; etc.;
3. Indicate the position of the branches on the main chain, numbering from the carboxyl group as carbon 1.

    Use prefixes if there is more than one side chain of the same type (e.g. di- is used if there are two of the same type, tri- if there are three, etc.)

Example:

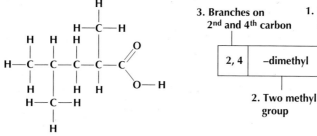

Drawing structural formulae for branched-chain carboxylic acids given the systematic name

1. identify the number of carbon atoms in the longest straight chain.
2. identify the branches and the numbers of the carbon atoms to which they are attached.

Example: The structural formula for 3-ethyl-2-methylhexanoic acid.

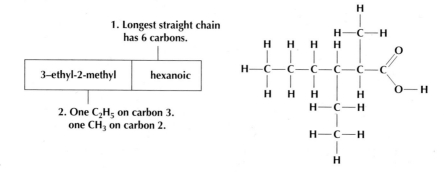

Reactions of carboxylic acids

Reduction

A carboxylic acid can be reduced to the corresponding aldehyde and alcohol. The oxygen:hydrogen ratio decreases from acid to alcohol, indicating reduction.

Example:

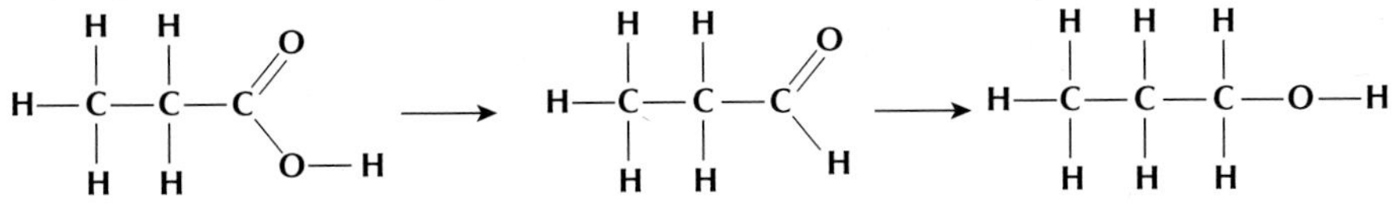

propanoic acid	propanal	propan-1-ol
O : H ratio : 1 : 3	1 : 6	1 : 8
(0·37)	(0·17)	(0·13)

Neutralisation

Carboxylic acids neutralise bases to form salts.

Ethanoic acid will react with calcium carbonate to form calcium ethanoate.

TOP TIP

Carboxylic acids show typical acid properties, and undergo neutralisation like common lab acids, such as hydrochloric acid.

$$2CH_3COOH(aq) + CaCO_3(s) \rightarrow (CH_3COO^-)_2Ca^{2+}\ (aq) + CO_2(g) + H_2O(\ell)$$

calcium ethanoate

Salts of carboxylic acids can be used as preservatives in foods. Calcium propanoate (E282) is used to stop the growth of mould, which can spoil foods. Sodium benzoate (E211) does a similar job in bottled sauces and soft drinks.

Quick Test 21

1. State the systematic name for the carboxylic acid shown:

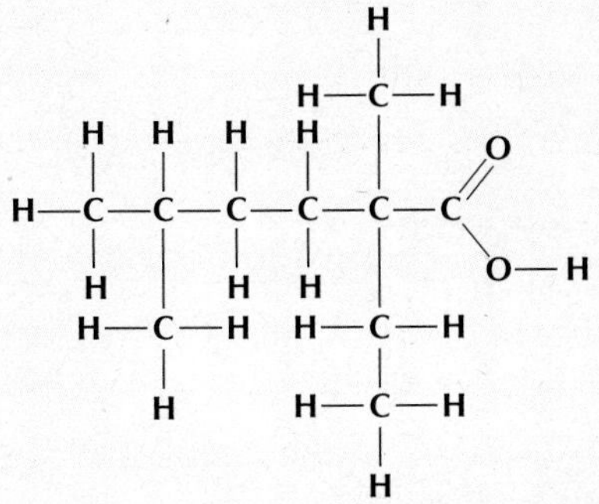

2. Draw the structural formula for 4-ethyl-3-methylheptanoic acid.

3. The conversion of ethanoic acid to ethanol is an example of reduction.
    Use oxygen:hydrogen ratios to explain this.

4. (a) Write a formula equation for propanoic acid reacting with magnesium carbonate.

    (b) Name the salt formed in part (a).

Oxidation of food 4

Antioxidants

Oxygen in the air can react with the chemicals in foods, causing them to spoil. Oils are particularly susceptible to oxidation. Many products (including free fatty acids) are formed, some of which have an unpleasant smell and taste. These fatty acids can be further oxidised at the double bonds, releasing volatile aldehydes which also smell. When an oil or fat is oxidised in this way, it is often said to be rancid.

To reduce the oxidation of chemicals in food, **antioxidants** are added to foodstuffs. If used properly, they can extend the shelf life of these products considerably.

Some antioxidants react with oxygen in preference to the foodstuff, and others prevent the formation of soapy-tasting peroxides.

A common antioxidant used in foodstuffs is ascorbic acid (vitamin C). The hydroxyl groups attached to the two carbons in the ring are oxidised to carbonyl groups.

O : H ratio : 6 : 8 (3 : 4 = 0·75)   6 : 6 (3 : 3 = 1)

The ion–electron equation and the oxygen:hydrogen ratio both indicate that the process is oxidation.

Isomerism

Isomers are molecules that have the same molecular formula, but have a different arrangement of the atoms in space.

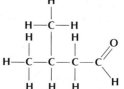

TOP TIP

Esters, aldehydes, ketones, alcohols and carboxylic acids can all exhibit some type of isomerism.

Chain isomerism

In chain isomerism, the atoms are arranged in a completely different order, which results in branches.

Example 1: Aldehydes with the molecular formula $C_5H_{10}O$:

Straight chain   2-methylbutanal   3-methylbutanal

Branched chains

Example 2: Carboxylic acids with the molecular formula $C_5H_{10}O_2$:

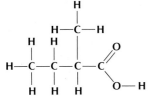

2-methylbutanoic acid   3-methylbutanoic acid

Position isomerism

In position isomerism, the basic carbon skeleton remains unchanged, but functional groups can be moved around on that skeleton.

Example 3: Ketones with the molecular formula $C_7H_{14}O$:

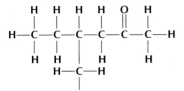

4-methylhexan-2-one   4-methylhexan-3-one

> **TOP TIP**
>
> Aldehydes and carboxylic acids can't have positional isomers, because the functional group must always be on the end carbon.

Example 4: Alcohols with the molecular formula $C_6H_{14}O$:

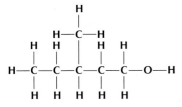

3-methylpentan-1-ol   3-methylpentan-2-ol

> **TOP TIP**
>
> Ketones and alcohols can also have chain isomers, i.e. branches in different positions.

Functional group isomerism

Functional group isomers contain different functional groups – that is, they belong to different families of compounds (different homologous series).

Example 5: Propanoic acid and methyl ethanoate (molecular formula: $C_3H_6O_2$:

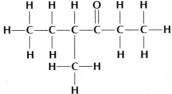

propanoic acid   methyl ethanoate

> **TOP TIP**
>
> If you are asked to draw the isomers from a given molecular formula, don't forget to think about all the possibilities. Can you branch the carbon chain? Can you move a functional group around on that chain? Is it possible to make more than one type of compound?

Quick Test 22

1. Draw aldehyde chain isomers with the molecular formula C_4H_8O and name them.
2. Draw aldeyde/ketone functional group isomers with the molecular formula C_3H_6O and name them.

Soaps, detergents and emulsions 1

Soap

Fats and oils are esters. When they are hydrolysed using alkali, salts of the long-chain fatty acids are formed. These are **soaps**. Glycerol is also formed and can be used in the food and pharmaceutical industries.

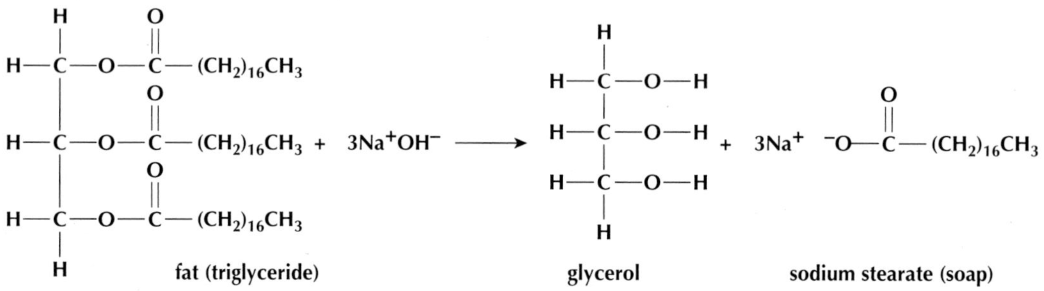

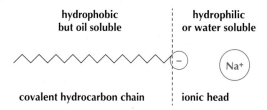

$$\begin{array}{ccc} \text{fat (triglyceride)} & \text{glycerol} & \text{sodium stearate (soap)} \end{array}$$

Structure of soap 'molecules'

A soap molecule has a long covalent hydrocarbon part, often referred to as the tail, and an ionic part, often referred to as the head.

hydrophobic
but oil soluble | hydrophilic
or water soluble

covalent hydrocarbon chain | ionic head

TOP TIP

Although often referred to as a 'molecule', soap is a salt and has ionic and covalent parts.

The tail, being covalent and non-polar, will be insoluble in water. It is termed **hydrophobic**, which means '**water-hating**'. It is, however, soluble in covalent substances like grease and oil.

The negatively charged head of the soap ion is termed **hydrophilic**, which means '**water-loving**'.

Cleansing action of soap

When soap is used to clean oil or grease from a surface, the hydrophobic tails of the soap ions bury into the oil and grease. Agitation causes small grease droplets (micelles) to form in the water. The negative charges on the heads repel each other and prevent the globules of oil from recombining. This allows the oil or grease to be washed off the surface.

(a)
soap dissolves in water, forming soap ions

(b)
tails of soap ions bury into grease, leaving heads in water

(c)
agitation begins to separate grease from surface

(d)
process continues, forming grease globules (miscelles) in the water

(e)
micelles are prevented from recombining by negative charges on heads of soap ions in the water

soap ions

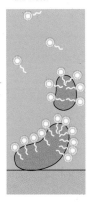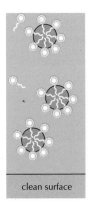

grease

surface   clean surface

One problem with soap is that it does not lather well in hard-water areas. Hard water contains calcium and magnesium ions. Insoluble salts like calcium stearate, known as scum, form. Detergents have been developed which do not form a scum.

Detergents

Detergents are cleaning agents which work in a similar way to soaps, but have the advantage of not forming insoluble salts in hard water. This means they can form a lather. Alkylbenzenesulfonates such as sodium dodecylbenzenesulfonate are detergents. They have a similar structure to soaps, in that they have a hydrophobic hydrocarbon tail and a hydrophilic ionic head. As a result, they clean in the same way as soaps.

sodium dodecylbenzenesulfonate: $CH_3(CH_2)_{11}C_6H_4$ $SO_3^- Na^+$

                     hydrocarbon tail   ionic head

The manufacture of detergents as cleansing agents is lucrative business all over the world.

The most common way of making dodecylbenzenesulfonate is by reacting benzene, C_6H_6, (from the oil industry) with dodecene ($C_{12}H_{24}$) and sulfur trioxide to give the sulfonic acid. The sulfonic acid is then neutralised with sodium hydroxide.

Quick Test 23

1. Explain why soap molecules are also called salts.

2. Describe the cleansing action of a soap molecule.

3. Explain why detergents can form a lather in hard-water areas.

Soaps, detergents and emulsions 2

Emulsions

An **emulsion** is a mixture of two or more liquids that are normally immiscible. The emulsion contains small droplets of one liquid dispersed in another, with the aid of an **emulsifier**. Emulsifiers allow normally immiscible liquids to mix and stop them separating.

Emulsions in food are mixtures of a fat/oil and water. Milk and mayonnaise are examples of emulsions. The natural emulsifier in milk is a protein called casein.

Emulsifiers are soap-like molecules that can be made by reacting edible oils with glycerol. One or two fatty acid groups are attached to the glycerol backbone, rather than the three in fats/oils (see Fats and oils page 40–41). This results in the emulsifier having molecules with a hydrophobic part (the fatty acid 'tail'), and a hydrophilic part – one or two hydroxyl groups or other water-soluble parts.

Diglyceride

hydrophobic
$$CH_3(CH_2)_{14}C(O)O—CH_2$$
$$CH_3(CH_2)_{14}C(O)O—CH$$
$$CH_2—O—P$$
$$CH_2$$
$$CH_2$$
$$^+N$$
$$CH_3$$
$$CH_3$$
$$CH_3$$

hydrophilic

Lecithin

hydrophilic
$$HO—CH_2$$
$$HO—CH$$
$$H_2C$$
$$O$$
$$C$$
$$O$$
$$CH_2 \ CH_2 \ CH_2 \ CH_2 \ CH_2 \ CH_2 \ CH_2 \ CH_2$$
$$CH_2 \ CH_2 \ CH_2 \ CH_2 \ CH_2 \ CH_2 \ CH_2 \ CH_2 \ CH_3$$

hydrophobic

In the emulsion, the hydrophobic part of the emulsifier dissolves in the fat/oil, and the hydrophilic part dissolves in the water. This holds the mixture together.

The inclusion of food additives is indicated on food labels by E numbers. The emulsifier known as E471 is the most commonly used emulsifier in the food industry. It contains a mixture of mono- and di-glycerides. Lecithin, found naturally in egg yolks, acts as the emulsifier in mayonnaise. The table on the next page gives some examples of foods that contain common emulsifiers.

Emulsifier	E number	Food product
lecithin	322	chocolate, margarine, chewing gum
mono-and di- glycerides	471	ice cream, bread, sausages

Adding mono- and di-glycerides to breads and cakes improves the texture and volume of the products. They keep cakes and bread soft, improving shelf life.

Emulsifiers help maintain their quality and freshness. In low-fat spreads, emulsifiers help prevent the growth of moulds, which would happen if the oil and fat separated.

Processed meats like salami contain mono- and di-glycerides as emulsifiers to ensure the even distribution of fat.

TOP TIP

The structures of emulsifiers are complicated. You are not expected to memorise them, but you should be able to recognise their structure. It is similar to fats and oils, but at least one of the fatty acids in the structure has been replaced by a hydrophilic group, such as –OH.

BERTOLLI LIGHT SPREAD MADE • W

38% VEGETABLE FAT SPREAD WITH 16% OLIVE OIL

Ingredients: Water, Olive Oil composed of Refined Olive Oils and Virgin Olive Oils (16%), Vegetable Oil, Modified Starch, Salt (0.8%), Buttermilk, Emulsifiers (Mono-and Di-glycerides of Fatty Acids, Sunflower Lecithin), Preservative (Potassium Sorbate), Citric Acid, Flavourings, Colour (Beta-Carotene), Vitamin A and D.

Quick Test 24

1. Polysorbate 80 is used as an emulsifier in ice cream. Describe the properties that polysorbate 80 is likely to have that make it useful as an emulsifier.

Fragrances

Essential oils

Fragrances are pleasant, sweet-smelling liquids, derived from **essential oils** in plants. Essential oils get their name from the word 'essence' and are the concentrated extracts of the volatile non-water soluble aroma compounds found in plants. They are widely used in perfumes, cosmetic products, cleaning products, and even as flavourings in food.

Terpenes

TOP TIP

You don't need to memorise the names and structures of terpenes, but you should be able to recognise a terpene from a given structural formula.

Essential oils are composed mainly of a class of compounds known as **terpenes**.

Terpenes are molecules that can be viewed as being based on isoprene units, 2-methylbuta-1,3-diene joined together.

Isoprene is a gaseous hydrocarbon which is naturally produced by different plants and emitted through their leaves into the atmosphere. Plants use isoprene to create terpenes by a complex biochemical process.

isoprene (2-methylbuta-1,3-diene)

The term **terpene** is often used to include terpene derivatives such as terpene alcohols, or aldehydes and ketones based on terpenes. The name ending normally indicates the functional group present. For example, -ene will indicate it is a hydrocarbon; -ol will indicate it contains a hydroxyl functional group, etc.

Geraniol can be extracted from the stems of the geranium plant

myrcene

geraniol

Myrcene and geraniol are **linear** terpenes – they are open chains.

Limonene is an example of a cyclic terpene.

Terpenes are classified according to the number of isoprene units that are joined together. The main classes of terpene in the essential oil plants are shown in the following table.

limonene

Class of terpene	Number of isoprene units	Formula	Example
monoterpene	2	$C_{10}H_{16}$	myrcene
sesquiterpene	3	$C_{15}H_{24}$	humulene
diterpene	4	$C_{20}H_{32}$	neocembrene

Oxidation of terpenes

Terpenes can oxidise easily. Oxidation of terpenes in plants is often responsible for the distinctive aroma of spices. Peppermint oil, for example, contains the terpene menthol as well as its oxidation product menthone. Both contribute to the peppermint flavour of the oil.

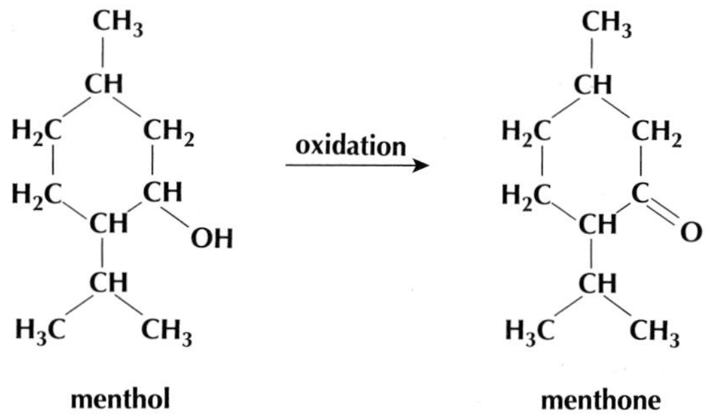

menthol → menthone

Quick Test 25

1. The structure of the terpene farnesol is shown below:

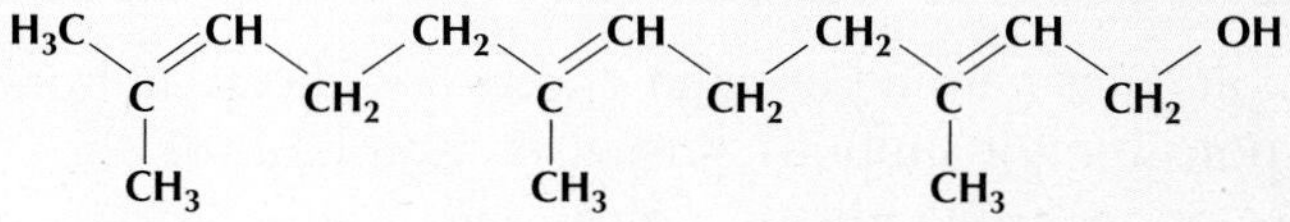

    (a) State which class of terpene farnesol belongs to.

    (b) Circle one isoprene unit in farnesol.

2. Limonene is a terpene found in orange juice which forms carvone if left exposed to the air:

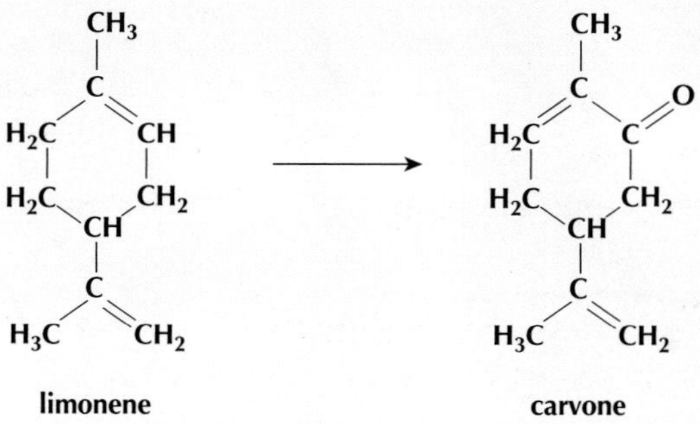

    limonene   carvone

    (a) State the type of reaction undergone by limonene.

    (b) An alcohol called carveol is also formed. Draw a possible structural formula for carveol.

Skin care 1

The damaging effect of ultraviolet (UV) radiation on skin

Sunlight is composed of many types of light, including invisible **ultraviolet (UV) light**. UV is high-energy radiation which causes many chemical reactions in our skin, including the production of vitamin D, tanning, burning and ageing. To prevent sunburn, **sun blocks** have been developed. They contain compounds like titanium dioxide, which reflect the UV light and stop it reaching the skin. Sun screens contain chemicals which act as UV filters and reduce the amount of UV reaching the skin.

Sun creams contain sunblock to reflect harmful UV rays

Excessive exposure to the sun can cause you to look much older. The sun's rays change the skin's texture by weakening its elasticity. This causes the skin to sag and appear leathery.

Free radical reactions

Exposure to UV light can result in molecules gaining enough energy for chemical bonds to be broken. This is the process responsible for sunburn, and also contributes to the ageing of skin. When UV light breaks bonds **free radicals** are formed. Free radicals have unpaired electrons, and so are very reactive. Free radicals will react quickly with neighbouring atoms to form new bonds. This gives rise to a **free radical chain reaction**.

There are three stages in a free radical chain reaction: initiation, propagation and termination.

Initiation: The first step, where free radicals are formed when a molecule absorbs radiation.

Propagation: Steps in which free radicals react, forming further free radicals that can themselves react.

Termination: Steps in which free radicals combine, slowing the rate and stopping the reaction.

> **TOP TIP**
>
> Reactions which are initiated by light are known as photochemical reactions.

> **TOP TIP**
>
> Free radicals have no charge and are represented as $X\bullet$ (g).

Example

The reaction of chlorine with hydrogen

Hydrogen and chlorine do not react until a bright light is shone at them. When the light is shone at them the gases react explosively.

Initiation: The UV light splits the chlorine molecule into two chlorine free radicals:

$$Cl_2(g) \quad \rightarrow \quad 2Cl\bullet(g)$$

Propagation: The chlorine radicals react with hydrogen molecules:

$$Cl\bullet(g) \quad + \quad H_2(g) \quad \rightarrow \quad HCl(g) \quad + \quad H\bullet(g)$$

The hydrogen free radical formed can react with a chlorine molecule:

$$H\bullet(g) \quad + \quad Cl_2(g) \quad \rightarrow \quad HCl(g) \quad + \quad Cl\bullet(g)$$

Termination: The reaction continues until all the molecules have been used up and the remaining free radicals have combined together:

$$H\bullet(g) \quad + \quad Cl\bullet(g) \quad \rightarrow \quad HCl(g)$$

TOP TIP

The propagation part of the reaction must have at least two steps. The termination step in any free radical chain reaction occurs when any of the radicals combine and no new radicals are formed.

Quick Test 26

1. Chlorine-containing compounds which reach the upper atmosphere are known to produce radicals which are destroying the ozone (O_3) layer.

    The reactions listed below are known to take place, but not in the order shown:

    (i) $ClO\cdot + O\cdot \quad \rightarrow \quad Cl\cdot + O_2$

    (ii) $CF_3Cl \quad \rightarrow \quad CF_3\cdot + Cl\cdot$

    (iii) $CF_3\cdot + CF_3\cdot \quad \rightarrow \quad C_2F_6$

    (iv) $Cl\cdot + O_3 \quad \rightarrow \quad ClO\cdot + O_2$

    (a) State what kind of reaction is taking place overall.

    (b) Step (iv) is a propagation step. State what each of steps (i)–(iii) are.

    (c) Suggest where the energy to start the reaction comes from.

    (d) Suggest why the C–Cl bond breaks in preference to the C–F bond.

Skin care 2

Free radical scavengers

Our bodies produce chemicals which will react with free radicals that might otherwise cause damage to cells. These substances are known as **free radical scavengers**. Antioxidants are free radical scavengers. These substances work by reacting with and removing the free radicals, and stopping chain reactions which would otherwise damage the cells. The process doesn't completely prevent damage, and becomes less effective as we grow older.

Free radical scavengers in foods

Many foods contain natural antioxidants such as vitamins C and E. There is strong evidence to support the health benefits of a class of compounds called polyphenols, particularly in compounds known as flavonoids. The herbs parsley and thyme are good sources of the flavanoid luteolin.

Parsley　　　　　*Thyme*　　　　　**luteolin**

Fruits such as blueberries and raspberries are a source of antioxidants

Antioxidants are also added to certain foods to help prevent the food from spoiling.

Free radical scavengers in cosmetics

Many creams contain compounds derived from vitamin C and vitamin E, which are powerful free radical scavengers. However, the effectiveness of compounds derived from them is disputed. Recent research has focused on polyphenols. They have been found to be able to pass into the epidermis and dermis, which demonstrates their potential as active ingredients in anti-ageing cosmetic products.

Free radical scavengers in plastics

Free radicals created in the atmosphere by UV radiation can cause plastics to degrade. Plastics such as polypropene and LDPE (low density polyethene) can discolour and crack. Oxygen reacts with the carbons in the polymer chains, forming carbonyl groups and weakening the plastics. Free radical scavengers are added to the plastics as UV stabilisers during the manufacturing process.

One common free radical scavenger added to plastics is benzophenone. It will react with reactive oxygen species, preventing damage to the polymer chains.

The durability of plastics used to make children's playground furniture is improved by incorporating free radical scavengers into them during the manufacturing process.

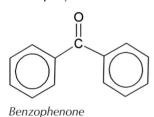

Benzophenone

Quick Test 27

1. Explain the need for free radical scavengers.
2. Give three examples of free radical scavengers and where they are used.

Learning checklist

Esters, fats and oils

In this section you have learned:

- An ester can be identified from the presence of the ester link:

- The ester name is in the form -yl oate where the first part of the name is derived from the parent alcohol and the second part from the parent carboxylic acid.
- Structural formulae for esters can be drawn given the name of the parent alcohol and parent carboxylic acid or given the name of the ester.
- Esters have characteristic smells, and are used as flavourings.
- Esters are also used as industrial solvents.
- Esters can be made by reacting alcohols with carboxylic acids.
- Ester formation, esterification, is an example of a condensation reaction.
- The ester link is formed when the hydroxyl group of the alcohol reacts with the carboxyl group of the acid, eliminating a small molecule of water in the process.
- Esters can be broken down (hydrolysed) into the parent alcohol and parent carboxylic acid.
- During hydrolysis, a molecule breaks down to two smaller molecules by reacting with water.
- Fats and oils are naturally occurring esters.
- The lower melting point of oils is related to the degree of unsaturation in the fatty acid molecules.
- Molecules in oils cannot pack as closely together as those in fats, leading to weaker intermolecular attractions and lower melting points.
- Fats and oils are an essential part of our diet, providing a concentrated energy source, and also transport for fat soluble vitamins in our bodies.

Proteins and enzymes

In this section you have learned:

- Proteins have many different functions in our bodies.
- Proteins are polypeptides made from amino acids.
- Amino acids contain an amino group ($-NH_2$) and a carboxyl group ($-COOH$).
- There are 20 different amino acids needed to make the proteins in our bodies.
- Ten of the amino acids must be obtained from dietary proteins, and these are referred to as essential amino acids.
- Amino acids join by a condensation reaction to form proteins.
- The link joining amino acid residues in polypeptide chains is a peptide (amide) link.

- Most enzymes are proteins and act as biological catalysts.
- During digestion, enzymes break proteins into smaller peptide chains and amino acids.
- The breakdown of proteins during digestion is an example of hydrolysis.
- The structural formulae of amino acids obtained on hydrolysis of protein can be identified from the protein structure.

Chemistry of cooking

In this section you have learned:
- Cooking causes proteins to denature.
- Many flavour molecules are aldehydes or ketones.
- Aldehydes and ketones contain the carbonyl functional group.
- The carbonyl group has the structure

$$\overset{\displaystyle O}{\underset{\displaystyle C}{\|}}$$

- In aldehydes the carbonyl functional group is on an end carbon.
- In ketones the carbonyl functional group is joined to two other carbons.
- When given the name for straight- and branched-chain aldehydes and ketones with up to eight carbons in the main chain, a structural formula can be drawn.
- When given the structure for straight- and branched-chain aldehydes and ketones with up to eight carbons in the main chain, it can be named.
- Aldehydes and ketones can be distinguished using acidified potassium dichromate, Fehling's solution or Tollens' reagent.
- The carbonyl functional group is polar, and leads to permanent dipole–permanent dipole attractions between molecules.
- The boiling points of aldehydes and ketones are lower than those of equivalent alcohols, which have hydrogen bonding.
- Low molecular mass aldehydes and ketones are water soluble due to hydrogen bonds forming between the carbonyl group and the water molecules.
- Low molecular mass aldehydes and ketones are very volatile.

Oxidation of food

In this section you have learned:
- How to name straight- and branched-chain aldehydes, ketones, alcohols and carboxylic acids.
- Alcohols can be classified as primary, secondary or tertiary, according to the number of carbon atoms that the carbon that is bonded to the hydroxyl group is attached to.
- Diols are alcohols with two hydroxyl groups and triols are alcohols with three hydroxyl groups.

- The ability to form hydrogen bonds affects properties of alcohols.
- Primary and secondary alcohols can be oxidised to aldehydes and ketones respectively. Tertiary alcohols cannot be oxidised.
- Aldehydes can be oxidised further to form carboxylic acids.
- Oxidation can be identified by an increase in the oxygen-to-hydrogen ratio, going from reactants to products.
- Carboxylic acids can be reduced to aldehydes, which can then be reduced to primary alcohols.
- Reduction can be identified by a decrease in the oxygen-to-hydrogen ratio, going from reactants to products.
- Carboxylic acids react with bases in neutralisation reactions to form salts.
- Oxidation is the major cause of food spoilage.
- Oxygen can react with the fatty acids in fats and oils, causing them to become rancid.
- Antioxidants are molecules which prevent other substances from oxidising.
- Ion–electron equations can be written for the oxidation of antioxidants.
- Aldehydes, ketones, alcohols and carboxylic acids exhibit isomerism.

Soaps, detergents and emulsions

In this section you have learned:
- Hydrolysis of fats and oils in alkaline conditions produces soaps.
- Soap molecules have an ionic head and a hydrocarbon tail.
- The ionic head is hydrophilic (water-loving) and dissolves well in water.
- The hydrocarbon tail is hydrophobic (water-hating) and dissolves well in grease and oil.
- How to explain the cleaning action of soap.
- Hard water contains calcium and magnesium ions which form a scum with soap.
- Detergents were developed to combat hard-water conditions.
- Detergents clean in the same way as soaps.
- Emulsions are formed when two normally immiscible liquids mix.
- Emulsifiers are compounds which allow two normally immiscible liquids to mix.
- Emulsifier molecules have both hydrophilic and hydrophobic parts in their structure.
- Emulsifiers are widely used in the food industry.

Fragrances

In this section you have learned:

- Fragrances can be due to essential oils in plants.
- Different methods can be used to extract oils from plants.
- Essential oils contain terpenes.
- Terpenes are molecules based on isoprene units joined together.
- The systematic name for isoprene is 2-methylbuta-1,3-diene.
- The term terpene is used to include compounds containing different functional groups.
- Oxidation of terpenes in plants is responsible for the aroma of many spices.

Skin care

In this section you have learned:

- UV light is high-energy radiation which can cause skin damage, including photoageing.
- Sunblocks can physically protect by reflecting UV radiation.
- Sunscreens can chemically protect by absorbing UV radiation.
- UV radiation causes harmful free radicals, which react with proteins in our skin to form wrinkles.
- Free radicals are single atoms or groups of atoms with unpaired electrons, and are very reactive.
- The formation of free radicals initiates chain reactions.
- The three stages of a free radical chain reaction are initiation, propagation and termination.
- Free radical scavengers react with free radicals, preventing free radical chain reactions taking place.
- Free radical scavengers are also known as antioxidants.
- Antioxidants are added to cosmetic products to prevent free radical damage to skin tissue.
- Antioxidants are added to foods to prevent them from spoiling and to improve shelf life.
- Free radical scavengers are added to plastics to slow down degradation by UV light and oxygen.

Getting the most from reactants 1

Maximising profit

When making new products, industrial chemists have to think about the cheapest and most efficient way of doing so in order to maximise profit.

Key considerations are:

- **Availability, sustainability** (long-term supply) and **cost** of **feedstocks** – the chemicals that have to be reacted or processed in some way in order to make a product.
- **Product yield** – the quantity of chemical produced by a particular chemical process.
- **Marketability of by-products** – often there are other products formed in addition to the intended product, which can be used elsewhere in the process or sold to other companies and used rather than being wasted.
- **Recycling reactants** – if unreacted chemicals can be fed back into the reactor, this improves the efficiency and profitability of the process as well as reducing potential environmental pollutants being released.
- **Energy requirements** – the cost of energy sources such as gas and electricity are steadily increasing, so low-energy processes are more profitable, and ways of using energy released during chemical reactions need to be found.

Case study – Ammonia and urea production in New Zealand

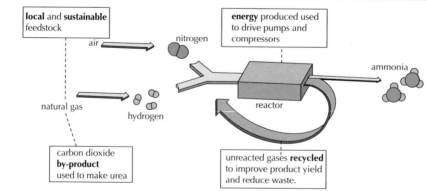

Ammonia and urea production in New Zealand, highlighting some of the ways the process is made as efficient as possible.

Minimising health risks and the impact on the environment

Industrial chemists have to think about health and safety issues to protect workers, the public, and the environment in general. In recent years the chemical industry has become more aware of **green chemistry** – the design of chemical products and processes that reduce or eliminate the use and production of hazardous substances.

The main principles of green chemistry are:

1. Waste prevention
2. More energy-efficient processes with fewer steps and fewer by-products improve the atom economy (see page 75)
3. Use of fewer hazardous and toxic chemicals
4. Safer processes and products
5. Design products which can be degraded in the environment
6. Use of more specific catalysts, such as enzymes which reduce energy requirements

Many pharmaceutical companies use enzyme catalysts to help achieve green chemistry ideals:

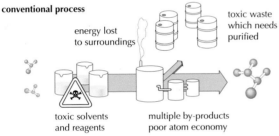

Many conventional industrial processes were inefficient and paid little attention to environmental effects.

> ### TOP TIP
>
> Although it is important that industrial chemists make products as cheaply as possible, they have to ensure that health and safety and environmental issues are addressed.

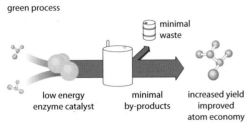

Green industrial processes both use and produce fewer toxic chemicals, and pay greater attention to the environment.

Quick Test 28

1. An industrial chemical manufacturer wants to adapt one of its processes so it becomes more 'green'. Outline the areas which need to be examined to achieve this.

Getting the most from reactants 2

Measuring the efficiency of a reaction

Manufacturers aim to use chemical reactions and processes that make the most effective use of available resources, whilst generating the smallest possible amount of waste material. The efficiency of a process can be measured by calculating the **percentage yield** and **atom economy**.

TOP TIP

Atom economy is calculated in the planning stage, but yield can only be found experimentally.

Percentage yield

Yield is the final, usable product of a process. Percentage yield compares the expected product quantity with the actual amount produced. The expected or theoretical yield is calculated from the balanced equation. You can find out how to calculate the theoretical yield on page 79.

$$\text{Percentage yield} = \frac{\text{actual yield}}{\text{theoretical yield}} \times 100$$

Example

Aspirin is a well-known painkiller made from salicylic acid:

$$\underset{\text{salicylic acid}}{C_7H_6O_3} \ + \ C_4H_6O_3 \ \rightarrow \ \underset{\text{aspirin}}{C_9H_8O_4} \ + \ C_2H_4O_2$$

The theoretical yield of aspirin when 100 g of salicylic acid is reacted is 130·4 g. The actual yield is 121·2 g. Calculate the percentage yield of aspirin.

Worked answer:

$$\text{Percentage yield} = \frac{\text{actual yield}}{\text{theoretical yield}} \times 100$$

$$= \frac{121 \cdot 2}{130 \cdot 4} \times 100$$

$$= 92 \cdot 9\%$$

This is a very high yield, so – using this as an indicator – the process could be considered to be efficient.

Atom economy

The atom economy compares the proportion of reactant atoms that end up in a useful product compared to the number that end up as waste. This is one of the key ideas in green chemistry. Atom economy can be calculated using the following equation:

$$\text{Atom economy (\%)} = \frac{\text{mass of desired product}}{\text{total mass of reactants}} \times 100$$

Example

Calculate the atom economy for the production of aspirin in the above reaction:

$$C_7H_6O_3 \quad + \quad C_4H_6O_3 \quad \rightarrow \quad C_9H_8O_4 \quad + \quad C_2H_4O_2$$

| salicylic acid | | aspirin | ethanoic acid |

Formula mass: 138   102   180   60

Worked answer:

$$\text{Atom economy (\%)} = \frac{\text{mass of desired product}}{\text{total mass of reactants}} \times 100$$

$$= \frac{180}{240} \times 100$$

$$= 75\%$$

Using atom economy as a measure of reaction efficiency, this reaction would be considered high. If the by-products of a reaction can be used, the atom economy is improved, and therefore the overall efficiency can be considered to be even higher.

In this reaction the other product is ethanoic acid, which is a useful product and therefore not wasted, so the reaction atom economy could be considered to be 100%.

TOP TIP

Both percentage yield and atom economy have to be considered when working out the efficiency of a chemical process. A reaction which has a high yield may have a poor atom economy if there are a number of by-products, so is not as efficient as it might at first seem.

Calculating the cost

If the cost of the feedstocks and the percentage yield for a reaction are known, then the cost of the reactants used in the reaction can be calculated.

Example

In the worked example on page 74, the percentage yield of aspirin from salicylic acid was calculated as 92·9%. If the average cost of salicylic acid is £4·90/kg, calculate the cost of the salicylic acid to produce 100 g of aspirin.

$$C_7H_6O_3 + C_4H_6O_3 \rightarrow C_9H_8O_4 + C_2H_4O_2$$

Formula mass: 138   102   180   60

Worked answer:

Step 1: Calculate the mass of salicylic acid needed to produce 1 g of aspirin if the reaction is 100% efficient.

	$C_7H_6O_3$ + $C_4H_6O_3$	$\rightarrow$	$C_9H_8O_4$ + $C_2H_4O_2$
Key relationship	1 mol	$\rightarrow$	1 mol
Convert to gram formula mass (GFM)	138 g	$\rightarrow$	180 g
So	$\dfrac{138}{180}$ g	$\rightarrow$	1 g
So	$\left(\dfrac{138}{180}\right) \times 100$	$\rightarrow$	100 g

So theoretically 76·7 g of salicylic acid produces 100 g of aspirin.

Step 2: Calculate the mass of reactant needed given the % yield of the reaction.

The percentage yield is 92·9 %, so the mass of salicylic acid needed to produce 100 g of aspirin is $\dfrac{100}{92·9} \times 76·7 = 82·6$ g

Step 3: Calculate the actual cost of the reactant.

1000 g of salicylic acid costs £4·90, so 82·6 g costs $\dfrac{82·6}{1000} \times £4·90 = £0·40$.

Why the yield is not 100%

There are several reasons:

1. The reaction may not go to completion: some of the reactants are not converted into products

2. Side-reactions, that produce by-products, may occur

3. Difficulty in purifying the product which can lead to loss of product

Quick Test 29

1. Ammonia is made industrially by the Haber process: $N_2 + 3H_2 \rightleftharpoons 2NH_3$

    When 112 g of nitrogen is reacted with hydrogen, the theoretical yield of ammonia is 136 g. The actual yield is 40·8 g.

    (a) Calculate the percentage yield of ammonia.

    (b) What is the atom economy for the reaction?

    (c) Comment on the overall efficiency of the reaction.

2. Phenol (C_6H_5OH) is an important industrial chemical which can be made from benzene (C_6H_6) as follows:

    $$C_6H_6 + H_2SO_4 + 2NaOH \rightarrow C_6H_5OH + Na_2SO_3 + 2H_2O$$

    78 g of benzene should theoretically produce 94 g of phenol. The actual yield is 75 g.

    (a) Calculate the percentage yield of phenol.

    (b) Calculate the atom economy for the production of phenol.

    (c) Comment on the overall efficiency of the reaction.

3. The percentage yield of iron produced from iron ore is 79%. If the average cost of iron ore (Fe_2O_3) is £83 per tonne, calculate the cost of the iron ore needed to produce each tonne of iron (to the nearest £). (1 tonne = 1000 kg.)

    $$\text{Balanced equation: } Fe_2O_3 + 3CO \rightarrow 2Fe + 3CO_2$$

Getting the most from reactants 3

Mole ratios in a formula

One mole of any substance contains the same number of **atoms** or **molecules** in an element as there are **molecules** or **ionic units** in a compound.

For example, 1 mole of aluminium has the same number of atoms as there are molecules in 1 mole of chlorine and 1 mole of carbon dioxide, and ionic units in 1 mole of sodium chloride.

One mole of molecules contains more than 1 mole of atoms.

For example, 1 mole of carbon dioxide (CO_2) contains 3 moles of atoms – 1 mole of carbon atoms and 2 moles of oxygen atoms. So, 0·5 moles of carbon dioxide contains 0·5 moles of carbon atoms and 1 mole of oxygen atoms, i.e. 1·5 moles of atoms.

One mole of ionic units has more than 1 mole of ions.

For example, 1 mole of sodium chloride (NaCl) contains 1 mole of Na^+ ions and 1 mole of Cl^- ions, i.e. 2 moles of ions in total. So, 0·5 moles of sodium chloride contains 0·5 moles of Na^+ ions and 0·5 moles of Cl^- ions, i.e. 1 mole of ions in total.

This relationship can be used to calculate masses of substances present in everyday foodstuffs.

Example

The nutritional information label on a packet of 'cup-a-soup' stated that it contained 0·55 g of sodium. The sodium is in the form of sodium chloride (NaCl).

Calculate the mass of sodium chloride in the packet of soup.

Worked answer:

Step 1: Calculate the number of moles of sodium ions – this is also the number of moles of NaCl ionic units.

GFM of Na = 23 g.

Number of moles of Na $= \dfrac{mass}{GFM}$

$= \dfrac{0.55}{23} = 0·024$ mol

The number of moles of NaCl is also 0·024.

GFM of NaCl = 23·0 + 35·5 = 58·5 g.

Step 2: Calculate the mass of NaCl from the number of moles.

Mass = moles × GFM

$= 0·024 × 58·5 = 1·40$ g

So, the mass of sodium chloride in the packet of soup is 1·40 g.

Nutrition information		
Made up as per instructions		
Typical values	**Per 100ml (as cosumed)**	**Per Sachet (as consumed)**
Energy	173kJ	387kJ
	(41kcal)	(92kcal)
Protein	0.5g	1.2g
Carbohydrates	7.2g	16.1g
of which sugar	1.4g	3.1g
Fat	1.1g	2.4g
of which saturates	0.9g	1.9g
Fibre	0.2g	0.5g
Sodium	0.25g	0.55g
	(250mg)	(550mg)

The nutritional information label from a soup packet showing the amount of sodium present.

Calculating masses reacting and produced

Quantities reacting and being produced can be calculated from balanced equations. We can do this because 1 mole of any substance contains the same number of atoms or molecules in an element, and molecules or ionic units in a compound.

The numbers in front of a formula in an equation tell us the number of **moles** reacting and being produced. The number 1 is never written in front of a formula in an equation – the fact that a formula is written tells you there must be at least 1 mole present.

The values obtained from the balanced equation are the theoretical values. The actual values are obtained by experiment.

> **TOP TIP**
>
> The theoretical yield calculated from the balanced equation is used along with the actual yield to calculate the percentage yield (see page 74).

Example

Calculate the theoretical mass of silver produced when 3·2 g of zinc metal is reacted with excess silver(I) nitrate solution.

Worked answer:

$$Zn \;+\; 2AgNO_3 \;\rightarrow\; 2\,Ag \;+\; Zn(NO_3)_2$$

	1 mol	2 mol	2 mol	1 mol

Key relationship   **1 mol** $\rightarrow$ **2 mol**

Converting to GFM: 65·4 g $\rightarrow$ 215·8 g (2 × 107·9)

So $1\cdot0 \text{ g} \left(\dfrac{65.4}{65.4}\right)$ $\rightarrow$ $3\cdot3 \text{ g} \left(\dfrac{215.8}{65.4}\right)$

So   3·2 g (1 × 3·2) $\rightarrow$ 7·82 g (3·3 × 3·2) = 10·56

The mass of silver produced is 10·56 g.

> **TOP TIP**
>
> Actual mass of a desired product in a chemical reaction will be less than the theoretical values because: by-products could form, the reaction may not go to completion, and reactants may not be pure.

> **TOP TIP**
>
> If the answer was asked for in moles instead of grams, convert to moles using: moles = $\dfrac{\text{mass}}{\text{GFM}}$

Quick Test 30

1. The nutritional information label on a family bag of potato crisps stated that it contained 0·82 g of sodium. The sodium is in the form of sodium chloride (NaCl). Calculate the mass of sodium chloride in the bag of potato crisps.

2. Calculate the mass of copper produced when 1·7 g of copper(I) sulfide is roasted in the air.

$$Cu_2S(s) \;+\; O_2(g) \;\rightarrow\; 2Cu(s) \;+\; SO_2(g)$$

Getting the most from reactants 4

Calculating the number of moles of solution reacting and produced from a balanced equation

TOP TIP

The relationship between concentration (c), moles (n) and volume (V) is given in the SQA data booklet as n = c × V.

Example

Calculate the number of **moles** of sodium hydroxide needed to completely neutralise 150 cm^3 of 0·1 mol l^{-1} sulfuric acid.

$$2NaOH \quad + \quad H_2SO_4 \quad \rightarrow \quad Na_2SO_4 \quad + \quad 2H_2O$$

Worked answer:

Work out the key mole relationship from the balanced equation.

$$2NaOH \quad + \quad H_2SO_4 \quad \rightarrow \quad Na_2SO_4 \quad + \quad 2H_2O$$

Key relationship **2 mol** **1 mol**

From the question, 150cm^3 of 0·1 mol l^{-1} sulfuric acid reacts completely.

Calculate number of moles: n = c × V

= 0·10 × 0·15

= 0·015 mol

The mole ratio shows that **twice** the number of moles of sodium hydroxide are needed to completely neutralise the sulfuric acid, i.e. **0·03** mol.

Calculating the concentration of a reactant from a balanced equation

Example

Calculate the concentration of a sodium hydroxide solution if 50 cm^3 of the solution were needed to neutralise 150 cm^3 of 0·1 mol l^{-1} sulfuric acid.

Worked answer:

Step 1: Calculate the number of moles of sulfuric acid reacting, and so the number of moles of sodium hydroxide needed, as done in the example above: **0·03** mol.

Step 2: Use the relationship $c = \dfrac{n}{V}$ to calculate the concentration of the sodium hydroxide:

$$c = \frac{n}{V}$$

So $\qquad\qquad c = \dfrac{0.03}{0.05} = \mathbf{0\cdot6} \text{ mol l}^{-1}$

Calculating the volume of solution needed for complete reaction from a balanced equation.

TOP TIP

If you are asked to 'show your working clearly' you must do so or you will lose marks. It is also helpful for you to lay out your working in a logical way, setting out each step. This also helps you to avoid putting the wrong figures into your calculator.

Example

Calculate the **volume** of $0\cdot1$ mol l^{-1} sodium hydroxide solution needed to neutralise 150 cm^3 of $0\cdot1$ mol l^{-1} sulfuric acid.

$$2NaOH + H_2SO_4 \rightarrow Na_2SO_4 + 2H_2O$$

Worked answer:

Step 1: Calculate the number of moles of sulfuric acid reacting, and so the number of moles of sodium hydroxide needed, as done in example 1 above: **$0\cdot03$** mol.

Step 2: Calculate the volume from the number of moles.

$$V = \frac{n}{c}$$

So $V = \dfrac{0.03}{0.1} = \mathbf{0\cdot30}$ l (300 cm^3).

Quick Test 31

1. Calculate the number of **moles** of sodium hydroxide solution needed to neutralise 20 cm^3 of $0\cdot1$ mol l^{-1} sulfuric acid.

$$2NaOH + H_2SO_4 \rightarrow Na_2SO_4 + 2H_2O$$

2. $17\cdot8$ cm^3 of potassium hydroxide solution is required to completely neutralise 40 cm^3 of $0\cdot1$ mol l^{-1} hydrochloric acid. Calculate the **concentration** of the potassium hydroxide solution.

$$KOH + HCl \rightarrow KCl + H_2O$$

3. Calculate the **volume** of $0\cdot25$ mol l^{-1} lithium hydroxide needed to completely neutralise 25 cm^3 of $0\cdot1$ mol l^{-1} of sulfuric acid.

$$2LiOH + H_2SO_4 \rightarrow Li_2SO_4 + 2H_2O$$

Getting the most from reactants 5

Reactants in excess

It is seldom the case that all of the reactants actually react to form products. One reactant is usually completely used up – this is the **limiting reactant**. Other reactants that are not completely used up are in **excess**. Calculations must be based on the quantity of the limiting reactant present.

Example

25 cm^3 of 0·2 mol l^{-1} potassium hydroxide solution is added to 15 cm^3 of 0·5 mol l^{-1} hydrochloric acid.

$$KOH + HCl \rightarrow KCl + H_2O$$

Show by calculation which reactant is in excess.

Worked answer:

Step 1: Work out the **key mole relationship** from the balanced equation.

$$KOH + HCl \rightarrow KCl + H_2O$$

Key relationship   **1 mol**   **1 mol**

Step 2: Calculate the **number of moles** of each reactant from the information in the question:

moles (n)   = concentration (c) × volume (V)

KOH: moles = 0·2 × 0·025

             = 0·0050 mol

HCl: moles = 0·5 × 0·015

             = 0·0075 mol

The potassium hydroxide and hydrochloric acid react in a 1:1 ratio, so the hydrochloric acid is in excess by 0·0025 mol (0·0075 – 0·0050) and the potassium hydroxide is the limiting reactant.

TOP TIP

The mole ratios from the balanced equation are key to calculating the number of moles of each chemical reacting.

Example

Hydrogen gas can be produced in the laboratory by reacting magnesium with hydrochloric acid. 1.6 g of magnesium was added to 250 cm³ of 0.5 mol l⁻¹ of hydrochloric acid.

Calculate which reactant is in excess.

$$Mg + 2HCl \rightarrow MgCl_2 + H_2$$

Worked answer:

Step 1: Work out the **key mole relationship** from the balanced equation.

$$Mg + 2HCl \rightarrow MgCl_2 + H_2$$
$$1\ mol \quad 2\ mol$$

Step 2: Calculate the **number of moles** of each reactant from the information in the question.

Magnesium: moles $= \dfrac{mass}{GFM}$

$= \dfrac{1 \cdot 6}{24 \cdot 3}$

$= 0 \cdot 066$ mol

Acid: moles $=$ concentration × volume

$= 0 \cdot 5 \times 0 \cdot 25$

$= 0 \cdot 125$ mol

TOP TIP

The volume of hydrogen gas produced can be calculated and this is covered in **Getting the most from reactants 7**

Because the mole ratio of magnesium:acid is 1:2, the number of moles of acid needed to react with all of the magnesium would be 2 × 0·066 = 0·132mol. There is only 0·125 mol of acid available, so the acid is the limiting reactant, and the magnesium is in excess by 0·0035 mol (0·0660 – 0·0625).

Quick Test 32

1. 1·1 g of calcium carbonate is added to 100 cm³ of 0·25 mol l⁻¹ hydrochloric acid.

$$CaCO_3 + 2HCl \rightarrow CaCl_2 + H_2O + CO_2$$

   (a) Calculate which reactant is in excess, and by how much.

   (b) Calculate the mass of calcium chloride produced.

2. 5·9 g of copper metal is added to 250 cm³ of 0·5 mol l⁻¹ of silver(I) nitrate solution.

$$Cu + 2AgNO_3 \rightarrow Cu(NO_3)_2 + 2Ag$$

   (a) Calculate which reactant is in excess, and by how much.

   (b) Calculate the mass of silver produced.

Getting the most from reactants 6

Molar volume

Molar volume (V_m) is the volume occupied by one mole of a gas when measured at a given temperature and pressure. This means that the molar volume changes depending on the temperature and pressure at which it is measured. The molar volume for **any** gas is the same so long as it is measured at the same temperature and pressure. Measured at standard temperature and pressure (s.t.p. = 0°C and 1 atmosphere pressure) molar volume is 22·4 l mol⁻¹. Measured at room temperature and pressure (20°C and 1 atmosphere pressure) molar volume is approximately 24·0 l mol⁻¹.

Calculating volumes of gas

If the molar volume is known then, the volume occupied by a quantity of gas can be calculated.

Example

Calculate the volume occupied by 0·3 mol of methane at room temperature and pressure.
$$V_m = 24·0 \text{ l mol}^{-1}$$

Worked answer:

$$1 \text{ mol} = 24·0 \text{ l mol}^{-1}$$
$$\text{So} \quad 0·3 \text{ mol} = 0·3 \times 24·0 = 7·2 \text{ l}$$

Example

Calculate the volume occupied by 2·3 g of carbon dioxide.
$$V_m = 23·8 \text{ l mol}^{-1}$$

Worked answer:

Step 1: Convert mass to moles.
$$\text{Moles} = \frac{\text{mass}}{\text{GFM}}$$
$$= 2·3/44·0$$
$$= 0·052 \text{ mol}$$

Step 2: Convert moles to volume.
$$1 \text{ mol} = 23·8 \text{ l mol}^{-1}$$
$$\text{So } 0·052 \text{ mol} = 0·052 \times 23·8 = 1·24 \text{ l}$$

Example

Calculate the mass of 6·3 l of hydrogen.

$$V_m = 23\cdot9 \text{ l mol}^{-1}$$

Worked answer:

Step 1: Calculate the **moles** of hydrogen.

23·9 l is the volume occupied by 1 mol of H_2

6·3 l will be the volume occupied by $\dfrac{6\cdot3 \times 1}{23\cdot9} = 0\cdot264$ mol

Step 2: Calculate the mass from the number of moles.

$$1 \text{ mol} = 2 \text{ g}$$

$$\text{So } 0\cdot264 \text{ mol} = 2 \times 0\cdot264 = 0\cdot528 \text{ g}$$

Quick Test 33

1. In each part of the question take the molar volume to be 23·7 l mol^{-1}.

    (a) Calculate the volume occupied by 0·15 mol of helium gas.

    (b) Calculate the volume occupied by 0·32 g of helium.

    (c) A weather balloon was filled with 58·4 l of helium gas. Calculate the mass of the helium in the balloon.

Getting the most from reactants 7

Volumes reacting and being produced

Volumes of gas reacting and being produced can be calculated if the mole ratios of reactants and products from the balanced equation are known.

Example

Calculate the volume of carbon dioxide gas produced when 250 cm³ of propane (C_3H_8) burns in excess oxygen.

$$C_3H_8 \text{ (g)} + 5O_2\text{(g)} \rightarrow 3CO_2\text{(g)} + 4H_2O(\ell)$$

Worked answer:

Step 1: Work out the **key mole relationship** from the balanced equation.

$$C_3H_8 \text{ (g)} + 5O_2\text{(g)} \rightarrow 3CO_2\text{(g)} + 4H_2O(\ell)$$

Key relationship   1 mol   3 mol

Step 2: Use the relationship 1 mol = V_m. This means that if one gas volume is known, and the mole ratio is known, then the other gas volumes can be calculated by proportion.

$$C_3H_8\text{(g)} + 5O_2\text{(g)} \rightarrow 3CO_2\text{(g)} + 4H_2O(\ell)$$

                        1 mol   3 mol

(vol is the volume of   1 vol   3 vol
gas given in the question)

            250 cm³   →   750 cm³

So 250 cm³ of methane produces 750 cm³ of carbon dioxide.

The question states that there is excess oxygen available, which means there is more than enough available to react with all of the propane. The exact volume of oxygen can be worked out from the mole ratio in the balanced equation – in this case 1:5, so the volume of oxygen used up is 1250 cm³.

Volumes of gas produced in reactions

When solids and solutions are involved in a reaction, it is possible to calculate the gas volumes produced.

You need to know the mole ratios from the balanced equation, and then calculate the number of moles actually being produced, lastly converting to volume.

Example

Calculate the volume of hydrogen gas produced when 0·75 g of magnesium reacts completely with excess hydrochloric acid. Take molar volume (V_m) to be 23·9 l mol^{-1}.

$$Mg + 2HCl \rightarrow MgCl_2 + H_2$$

Worked answer:

Step 1: Work out the key mole relationship from the balanced equation.

$$Mg + 2HCl \rightarrow MgCl_2 + H_2$$

$$1 \text{ mol} \rightarrow 1 \text{ mol}$$

$$1 \text{ mol} \rightarrow V_m (23·9 \text{ l mol}^{-1})$$

Step 2: Calculate the moles of magnesium reacting.

$$\text{Moles} = \frac{\text{mass}}{\text{GFM}}$$

$$= \frac{0·75}{24·3}$$

$$= 0·031 \text{ mol}$$

Step 3: Calculate the volume of gas produced from the mole:V_m ratio in step 2.

$$1 \text{ mol} \rightarrow V_m (23·9 \text{ l mol}^{-1})$$

$$\text{So } 0·031 \text{ mol} \rightarrow 0·031 \times 23·9 = \mathbf{0·74 \text{ l}}$$

Quick Test 34

1. Butane gas (C_4H_{10}) is used as a portable fuel. It burns to produce carbon dioxide and water.

    $$C_4H_{10} (g) + \frac{13}{2} O_2(g) \rightarrow 4CO_2(g) + 5H_2O(\ell)$$

    (a) Calculate the volume of carbon dioxide produced when 125 cm^3 of butane is burned in excess oxygen.

    (b) Calculate the volume of oxygen which would be left if 1 l of oxygen was mixed with 100 cm^3 of butane and the mixture ignited.

2. Look at Worked Example 2 in **Getting the most from reactants 5**. Calculate the volume of hydrogen gas produced when all of the acid reacts. (Take molar volume to be 23·9 l mol^{-1}.)

Equilibria 1

Reversible reactions and equilibrium

Many of the reactions we have looked at so far, such as neutralisation and combustion, appear to be irreversible and, for practical purposes, they are. Burning magnesium in oxygen can be considered as a reaction which goes to completion:

$$2Mg(s) + O_2(g) \rightarrow 2MgO(s)$$

The single arrow in the equation tells us the reaction is irreversible and goes to completion.

However, many reactions can be reversed. A **reversible reaction** is one which can take place in both directions, i.e. the products can re-form reactants.

A reversible reaction you are likely to have seen in the laboratory is the colour-change which happens when water is added to blue cobalt chloride paper and it turns pink. The paper changes back to blue when the paper is warmed and the water is evaporated off.

Any general reversible reaction can be represented as:

$$A + B \rightleftharpoons C + D$$

The double arrow indicates the reaction is reversible. At the start: Concentrations of A and B are high, so rate of forward reaction (left to right) is high.

Concentrations of C and D are zero, so rate of backward reaction (right to left) is zero.

As the reaction progresses: Concentrations of A and B decrease, so rate of forward reaction decreases.

Concentrations of C and D increase, so rate of backward reaction increases.

This continues until the two rates are equal – a state of balance is formed. This is known as **dynamic equilibrium**. At this point the concentrations of reactants and products are unchanging, but seldom the same.

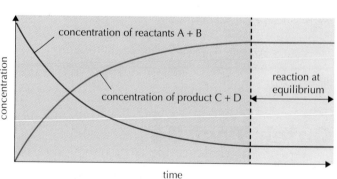

The changes in concentration of reactants and products and rate of the forward and backward reactions with time are shown in the graphs.

The position of equilibrium

In some reactions equilibrium isn't reached until the forward reaction is nearly complete – we say the equilibrium lies to the right. The concentration of the products is greater than the concentration of the reactants. This is shown in the left-hand graph, for the general reaction $A + B \rightleftharpoons C + D$.

In other reactions, equilibrium can be reached early in the reaction – we say the equilibrium lies to the left. This is shown in the right-hand graph. The concentrations of C and D are much less than A and B so the equilibrium is to the left.

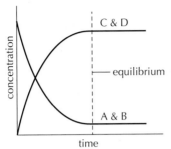 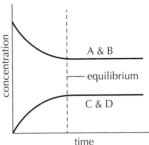

For a reversible reaction which forms an equilibrium, the same equilibrium position is reached regardless of whether we start from reactants or products.

Quick Test 35

1. A flask (flask 1) containing 0·5 mol of hydrogen and 0·5 mol of iodine, and another flask (flask 2) with 1 mol of hydrogen iodide were placed in a water bath and kept at the same temperature. After 90 mins the concentration of hydrogen iodide in both flasks was found to be 0·78 mol.

    (a) Explain this result.

    (b) Write a balanced formula equation to summarise the reaction.

2. The graph shows how the concentration of reactants and products changes as an equilibrium is formed.

    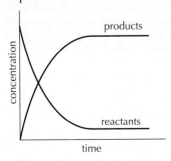

    (a) (i) Does the graph represent the change in concentration for an equilibrium which lies to the left or right?

        (ii) Explain your answer.

    (b) At which point on the graph is equilibrium reached?

Equilibria 2

Changing the position of equilibrium

Many industrial processes involve reactions which are reversible. This creates challenges for industrial chemists, who want to make as much product as possible, as economically as possible. They have to be able to change the reaction conditions so that the maximum amount of product is formed. Equilibrium is reached when the rate of the forward and backward reactions become equal, so the factors which affect the rate of a reaction might effect the equilibrium position.

TOP TIP

The position of the equilibrium of a system changes to minimise the effect of any imposed change in condition – concentration, pressure and temperature.

Changing concentration

The reaction between iron(III) ions (Fe^{3+}) and thiocyanate ions (CNS^-) can be used to show the effect on an equilibrium when the concentration of reactants or products are changed:

$$Fe^{3+}(aq) \quad + \quad CNS^-(aq) \quad \rightleftharpoons \quad [FeCNS]^{2+}(aq)$$

pale yellow   colourless   red

The two reactants are mixed and a deep red solution forms, which is diluted with water until it is an orange colour, and finally poured into four separate test tubes. Test tube A is kept as a control – nothing else is added to it. The other three solutions (B, C and D) each have chemicals added to them, which causes a colour-change in each test tube, as shown in the diagram.

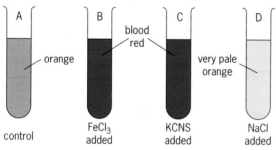

A — orange — control
B — blood red — FeCl₃ added
C — KCNS added
D — very pale orange — NaCl added

TOP TIP

Increasing the concentration of reactants shifts the equilibrium position right, i.e. more product is formed.
Reducing the concentration of reactants shifts the equilibrium position left, i.e. product re-forms reactants.
Removing product shifts the equilibrium position right, i.e. more product is formed.
Adding product shifts the equilibrium position left, i.e. more reactant is formed.

Observations:

Test tubes B and C: The concentration of one of the reactants is increased. The equilibrium position shifts to the right, forming more product. This is indicated by the blood-red colour forming as more $[FeCNS]^{2+}(aq)$ ions are made.

Test tube D: Adding NaCl removes $Fe^{3+}(aq)$ ions – the Cl^- ions form a complex with $Fe^{3+}(aq)$ ions. This results in the concentration of $Fe^{3+}(aq)$ ions decreasing so the equilibrium shifts left, forming more reactants. This is indicated by the orange solution becoming paler as the red $[FeCNS]^{2+}(aq)$ ions form $Fe^{3+}(aq)$ (pale yellow) and $CNS^-(aq)$ (colourless) ions.

Changing pressure

The effect of changing the pressure can be observed using nitrogen dioxide (NO_2), a brown gas. Nitrogen dioxide actually exists in equilibrium with dinitrogen tetroxide (N_2O_4), which is colourless.

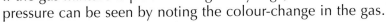

$$N_2O_4(g) \rightleftharpoons 2NO_2(g)$$
$$\text{colourless} \qquad \text{brown}$$

If the gas mixture is placed in a glass syringe, the effect of increasing and decreasing the pressure can be seen by noting the colour-change in the gas.

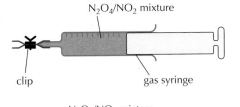

clip   gas syringe

N_2O_4/NO_2 mixture

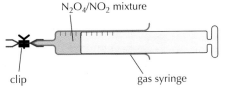

clip   gas syringe

More N_2O_4 formed - equilibrium shifts left in direction of fewer moles of molecules.

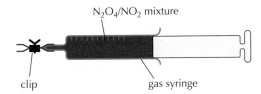

clip   gas syringe

More NO_2 formed - equilibrium shifts right in direction of more moles of molecules.

> **TOP TIP**
>
> Increasing pressure causes an equilibrium to move towards where there are fewer moles of gas molecules.
> Decreasing the pressure causes an equilibrium to move towards where there are more moles of gas molecules.
> If an equilibrium has the same number of moles of gas on each side of the equation then changing the pressure will have no effect on the position of the equilibrium.
> If an equilibrium has no gases then changing pressure will have no effect on the position of equilibrium.

Quick Test 36

1. Bromine dissolves in water to form the equilibrium shown in the following equation:

$$Br_2(\ell) + H_2O(\ell) \rightleftharpoons Br^-(aq) + BrO^-(aq) + 2H^+(aq)$$

(a) Explain the effect on the equilibrium if dilute nitric acid (source of $H^+(aq)$ ions) is added.

(b) Explain the effect on the equilibrium if dilute potassium hydroxide (source of OH^- (aq) ions) is added.

2. In the industrial preparation of methanol, a mixture of carbon monoxide and hydrogen gas (synthesis gas) is passed over a catalyst:

$$CO(g) + 2H_2(g) \rightleftharpoons CH_3OH(g)$$

(a) (i) State the pressure conditions which would most likely be used to improve the amount of methanol.

(ii) Justify your answer.

(b) The methanol is liquefied and drained off as it forms. Explain why this increases the yield of methanol.

Equilibria 3

Changing temperature

The effect of temperature on an equilibrium depends on whether a reaction is exothermic or endothermic. If, for example, the forward reaction is exothermic, then the backward reaction will be endothermic. The effect of changing temperature can be shown using the nitrogen dioxide (NO_2), dinitrogen tetroxide (N_2O_4) equilibrium:

$$N_2O_4 \rightleftharpoons 2NO_2 \qquad \Delta H = \text{positive}$$

colourless   brown

For an equation written this way, the ΔH value is for the forward reaction, which in this case is endothermic (positive). The enthalpy change for the backward reaction must therefore be exothermic (negative).

Three test tubes of the gas mixture are prepared so that they all have the same brown colour at room temperature. One is kept at room temperature (B), another, (C), is placed in a beaker of hot water to raise the temperature and the third, (A), is placed in a freezing ice/salt mixture to lower the temperature.

> **TOP TIP**
>
> For a system in equilibrium, a rise in temperature favours the endothermic reaction. A decrease in temperature favours the exothermic reaction.

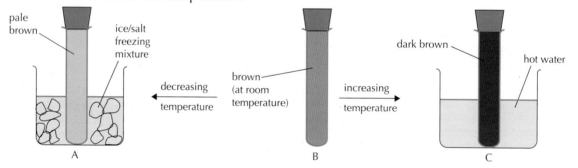

Observations:

Test tube C: The colour gradually darkens compared to (B), indicating that more NO_2 (brown) is formed. The equilibrium is moving right, the direction of the endothermic reaction.

Test tube A: The colour gradually lightens compared to (B), indicating that more (N_2O_4) (colourless) is formed. The equilibrium is moving left, the direction of the exothermic reaction.

Adding a catalyst

A catalyst speeds up a reaction because it gives a different route to products, which overall has a lower activation energy (E_a) than the reaction would have without a catalyst. This is shown in the potential energy diagram below.

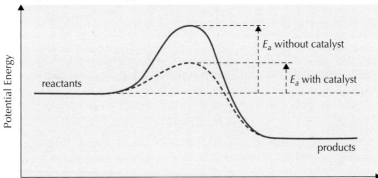

The potential energy diagram above shows that not only does a catalyst lower the energy of activation for the forward reaction, but also for the backward reaction. This means that in a reversible reaction, the rate of the backward reaction will also be increased. This means that the catalyst does not affect the equilibrium position – it does, however, allow the equilibrium position to be reached more quickly. Catalysts also allow reactions to be carried out at lower temperatures, which makes the reactions more economical.

> **TOP TIP**
>
> Catalysts are an essential in the economical production of chemicals on an industrial scale, many of which are produced by routes that involve reactions which are reversible.

> **TOP TIP**
>
> If a system at equilibrium is subjected to a change, the system will adjust to oppose the effect of the change.

Quick Test 37

1. In the industrial manufacture of methanol, the mixture of carbon monoxide and hydrogen gas (synthesis gas) used can be produced from methane:

    $$CH_4(g) \; + \; H_2O(g) \; \rightleftharpoons \; CO(g) \; + \; 3H_2(g) \quad \Delta H = +206 \text{ kJ mol}^{-1}$$

    (a) Predict the effect increasing the temperature would have on the yield of synthesis gas.

    (b) Explain the reasoning behind your answer.

2. (a) Predict the effect increasing the temperature would have on the following equilibrium:

    $$4NH_3(g) \; + \; 5O_2(g) \; \rightleftharpoons \; 4NO(g) \; + \; 6H_2O(\ell) \quad \Delta H = -908 \text{ kJ mol}^{-1}$$

    (b) Explain the reasoning behind your answer.

3. (a) Explain why the addition of a catalyst to a reversible reaction has no effect on the position of the equilibrium, even although it provides a route from reactants to products which has a lower activation energy.

    (b) What effect does the addition of a catalyst have on a chemical equilibrium?

Equilibria 4

Equilibria in industry

The Haber process is the industrial manufacture of ammonia from nitrogen and hydrogen:

$$N_2(g) + 3H_2(g) \rightleftharpoons 2NH_3(g) \quad \Delta H = -92 \text{ kJ mol}^{-1}$$

This is a reversible reaction, which in a closed system would form a dynamic equilibrium and thus, the equilibrium position can be changed by altering certain conditions.

1. Pressure

There are four moles of gases on the left-hand side and two moles on the right-hand side. Increasing the pressure would cause the equilibrium to shift to the right because there are fewer moles of gas on the right, i.e. more ammonia would be formed.

2. Temperature

The forward reaction is exothermic, so a drop in temperature would cause the equilibrium to shift to the right.

In industry, however, the conditions used are not always the theoretical conditions, mainly for economic reasons. In choosing the right temperature, manufacturers must consider the percentage yield and rate of reaction. Temperature has a major effect on the rate of reaction. If the temperature is too low, which would in theory result in more ammonia being formed, the rate at which ammonia is formed would not be fast enough. Manufacturers use a combination of a moderate temperature and a catalyst to ensure the rate of production of ammonia is high.

High pressure improves the yield and rate of formation, but the higher the pressure, the more it costs to build and run the plant. The higher the pressure, the thicker the pipes need to be, and the more energy is needed.

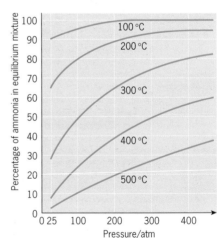

The graph shows how the percentage of ammonia in the equilibrium mixture varies ay different pressures and temperatures.

Most manufacturers use:

- a relatively high pressure of between 100 and 200 atmospheres;
- a moderate temperature of around 400°C;
- a finely divided iron catalyst.

The operating conditions are not a closed system. The ammonia is continually removed so equilibrium is not reached, and the reactants are recycled and continuously passed over the catalyst. This improves the yield of ammonia by up to 98%. Researchers are continuing to search for alternative catalysts which will work at lower temperature and pressure, reducing costs. In the USA scientists are working on alternative sources of hydrogen. The current source of hydrogen is mainly natural gas, which is a finite resource. Researchers are using electricity produced from wind power to electrolyse water and produce hydrogen.

The diagram opposite summarises the effect of increasing temperature and pressure in the Haber process.

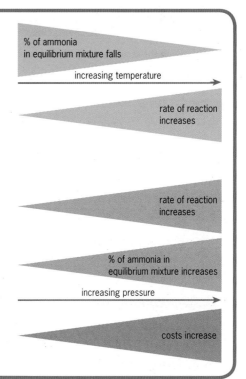

Quick Test 38

The flow diagram summarises the ammonia manufacturing process.

1. In the steam-reforming stage, hydrogen is made by reacting methane and steam at 850°C and 30 atmospheres pressure:

$$CH_4 (g) + H_2O (g) \rightleftharpoons CO (g) + 3H_2(g) \qquad \Delta H = +210KJ \ mol^{-1}$$

   (a) Explain why a high temperature is used.

   (b) Suggest why a high pressure is used, even though the equation indicates a low pressure should be used.

2. Ammonia is formed in the synthesis reaction vessel.

   (a) Explain why the iron catalyst is finely divided.

   (b) What is the effect of reducing the concentration of ammonia in the equilibrium mixture by removing it in the ammonia condenser?

   (c) Suggest why the $N_2:H_2$ mixture is in a 1:3 ratio as it enters the synthesis reaction vessel.

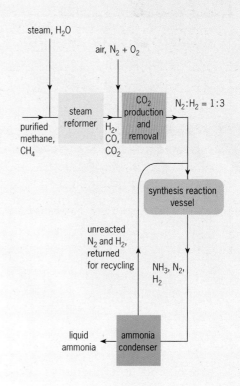

Chemical energy 1

Controlling the amount of heat

For industrial processes, chemists have to be able to calculate the amount of heat given out or taken in. Endothermic reactions require heat to be supplied and this is likely to push costs up. If a reaction is exothermic, the heat will have to be removed to prevent the temperature getting too high and causing thermal runaway.

Thermal runaway refers to a situation where an increase in temperature changes the conditions in a way that causes a further increase in temperature. This causes the reaction to go out of control, and can often result in an explosion.

Most industrial reactions are exothermic, and nearly every large-scale chemical process has at least one step with the potential for thermal runaway, if the conditions are right. This means they must have a way of allowing heat to escape.

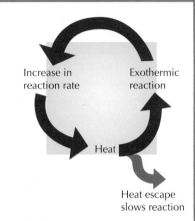

Increase in reaction rate

Exothermic reaction

Heat

Heat escape slows reaction

TOP TIP

Stored chemical energy is known as enthalpy. Remind yourself about enthalpy changes by looking at 'Controlling the rate of reaction 5: Reaction profiles'.

Measuring enthalpy changes by experiment

Enthalpy of combustion

At National 5, you learned how to calculate the enthalpy change when a substance burns completely in oxygen. This is important when working out, for example, how effective a fuel a certain substance might be. To make a fair comparison, the enthalpy change when burning the same quantity of fuel should be compared. Enthalpy changes given in data books are for **one mole** of a substance burning completely at standard temperature (25°C) and pressure (1 atmosphere), all substances being in their standard states; this is known as the **standard enthalpy of combustion (ΔH_c)**.

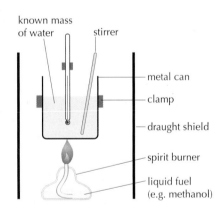

known mass of water

stirrer

metal can

clamp

draught shield

spirit burner

liquid fuel (e.g. methanol)

The diagram shows the arrangement which can be used to work out the enthalpy of combustion of methanol.

From the experimental results, the amount of energy transferred to the water (E_h) can be calculated by entering the values into a simple equation which connects them:

$$E_h = c \times m \times \Delta T$$

The change in temperature (ΔT) can be worked out by subtracting the final temperature from the initial temperature.

The mass of water (**m**), in kg, is known (1 cm^3 = 1 g).

The energy needed to raise the temperature of 1 kg (1 litre) of water by 1°C [the specific heat of water, (**c**)] is 4·18 kJ kg$^{-1°}$C^{-1}.

The mass of methanol burned is calculated by weighing the burner before and after the experiment.

Example

0·68 g of methanol raises the temperature of 100 g of water in the metal can by 22·4°C. Calculate the enthalpy of combustion (**ΔH$_c$**) of methanol.

Worked answer:

Step 1: Work out the energy transferred to the water (**E$_h$**):

$$E_h = c \times m \times \Delta T$$
$$= 4·18 \times 0·1 \times 22·4$$
$$E_h = 9·36 \text{ kJ}$$

> **TOP TIP**
>
> There is a – sign before the numerical value because the reaction is exothermic. The units are kJ mol^{-1} because enthalpy of combustion is the energy released when 1 mole of methanol is burned.

Step 2: Calculate the amount of fuel used, in moles:
  Mass of 1 mole of methanol (CH$_3$OH) = 32·0 g.
  Moles of methanol in 0·68 g = mass/mass of 1 mole = $\dfrac{0·68}{32·0}$ = 0·021 mol

Step 3: Work out the energy transferred to the water when 1 mole of methanol burns, i.e. the enthalpy of combustion:
  0·021 mole of methanol releases 9·36 kJ
  So 1 mole of methanol releases $\dfrac{9·36}{0·021}$ = 445·7 kJ mol^{-1}.
  So the enthalpy of combustion of methanol = –445·7 kJ mol^{-1}.

The standard enthalpy of combustion of methanol from the SQA data booklet is –726 kJ mol^{-1}. The experimental value for methanol is a fair bit lower than the standard value. This is because the experimental value is not obtained under standard conditions, and there are considerable heat losses to the surrounding area from the flame and the metal can. Therefore, not all the heat from the flame and the metal can reach the water.

> **TOP TIP**
>
> The relationship **E$_h$ = c × m × ΔT** is in the SQA data booklet, so you don't have to memorise it.

Quick Test 39

1. (a) 0·38 g of propan-1-ol (C$_3$H$_7$OH) raises the temperature of 100 g of water in a metal can by 25·4°C. Calculate the enthalpy of combustion of propan-1-ol

    (b) Suggest why the experimental results are less than the data booklet values.

2. 0·39 g of ethanol was burned and the heat produced used to heat 100 g of water. 8·33 kJ of energy was produced. Calculate the rise in temperature of the water, assuming all of the energy was transferred to the water.

Chemical energy 2

Enthalpy of reactions

An experiment can be carried out to collect data to calculate enthalpy change when an acid neutralises and alkali.

An example is the neutralisation reaction between hydrochloric acid and potassium hydroxide, which is an exothermic reaction:

hydrochloric acid + potassium hydroxide → water + potassium chloride

$HCl\ (aq)$ + $KOH(aq)$ → $H_2O(\ell)$ + $KCl\ (aq)$

The enthalpy of reaction can be determined in the laboratory using the simple calorimeter shown in the diagram opposite.

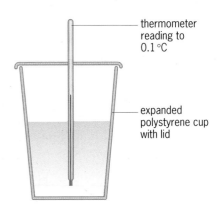

thermometer reading to 0.1 °C

expanded polystyrene cup with lid

The experiment is carried out as follows:

Step 1: Measure out separately known volumes of acid and alkali, measuring the temperature of each.

Step 2: Mix the acid with the alkali in a polystyrene cup, stir the mixture, and record the highest temperature reached.

Step 3: Calculate the change in temperature (ΔT) by subtracting the average temperature of the acid and alkali from the highest temperature recorded.

The enthalpy change can be calculated using $E_h = c \times m \times \Delta T$. In this experiment the mass (m) is the combined volume of acid and alkali. Even though it is not pure water absorbing the heat produced, the specific heat of water (c) is still used, i.e. 4·18 kJ $kg^{-1}\,{}^{\circ}C^{-1}$.

TOP TIP

When calculating the enthalpy change for any reaction, it is usually worked out for 1 mole of reactant or product.

Example

A group of students mixed 50 cm³ of 1·0 mol⁻¹ hydrochloric acid with 50 cm³ of 1·0 mol⁻¹ potassium hydroxide. The temperature rise was 6·3°C. Calculate the enthalpy of change when 1 mole of water is produced.

Worked answer:

Step 1: Work out the energy transferred to the water (E_h):
(m = 50 + 50 = 100 cm³ = 100 g = 0·1 kg)

$$E_h = c \times m \times \Delta T$$
$$= 4\cdot18 \times 0\cdot1 \times 6\cdot3$$
$$E_h = 2\cdot63 \text{ kJ}$$

Step 2: Calculate the number of moles of acid (or alkali) used:

$$\text{moles} = \text{concentration} \times \text{volume}$$
$$= 1\cdot0 \times 0\cdot05$$
$$= 0\cdot05 \text{ mol}$$

We know that the reaction taking place is:

$$H^+ (aq) + OH^- (aq) \rightarrow H_2O(\ell)$$

From the balanced equation:

| 1 mol | 1 mol | → | 1 mol |

So 0·05 mol   0·05 mol   0·05 mol
So 0·05 mol of water is formed.

Step 3: Work out the energy transferred when 1 mole of water is formed:
When 0.05 moles of water is formed, the energy released is 2·63 kJ

So, when 1 mole of water is formed, the enthalpy change is $\dfrac{2.63}{0.05}$
= −52·7 kj mol⁻¹.

> **TOP TIP**
>
> The actual value for the enthalpy change when 1 mole of water is formed in a neutralisation reaction is −57·2 kJ mol⁻¹. The difference in values is mainly due to heat loss to the surroundings during the laboratory experiment. The value is the same for any acid and alkali reacting because the same reaction is happening each time, i.e. water is produced.

Quick Test 40

1. A group of students mixed 25 cm³ of 1·0 mol⁻¹ hydrochloric acid with 25 cm³ of 1·0 mol⁻¹ sodium hydroxide. The temperature rise was 6·4°C.

$$HCl(aq) + NaOH(aq) \rightarrow H_2O(\ell) + NaCl(aq)$$

Calculate the enthalpy change when 1 mole of water is produced.

2. Explain why the enthalpy changes when 1 mol of water is formed in a neutralisation reaction is the same, no matter which acid or alkali is used.

Chemical energy 3

Enthalpy of formation

The **standard enthalpy of formation (ΔH_f)** is the enthalpy change when one mole of a compound is formed from its elements in their standard states, and under standard conditions (25°C and 1 atmosphere). An example is the formation of methane (CH_4) from its elements, carbon and hydrogen:

$$C(s) + 2H_2(g) \rightarrow CH_4(g) \qquad \Delta H_f = -75 \text{ kJ mol}^{-1}$$

Note that each element is in its standard state and one mole of product is formed.

The enthalpy of formation of any element is zero, as there is no change involved in the formation of an element. The importance of enthalpies of formation is that they can be used to find the enthalpy changes of other reactions.

TOP TIP

Enthalpies of formation can be found in the SQA data booklet.

Energy cycles and Hess's law

Some enthalpies of formation can be measured experimentally. However, many enthalpies of formation cannot be measured directly, so an indirect approach using energy cycles is used.

Using energy cycles to work out enthalpy of formation relies on **Hess's law**, which states that the enthalpy change for a chemical reaction is independent of the route taken from reactants to products, so long as the starting and finishing conditions are the same for each route. This is shown in an energy cycle in the diagram opposite:

The overall enthalpy change for route 2 must be the same as for route 1:

$$\Delta H_1 = \Delta H_2 + \Delta H_3 + \Delta H_4$$

We can use this idea to work out the enthalpy of formation of methane:

$$C(s) + 2H_2(g) \rightarrow CH_4(g) \ \Delta H_f \text{ (target equation)}$$

Methane cannot be made directly from its elements, carbon and hydrogen, but Hess's law can be used to calculate the enthalpy of formation by an indirect route. We can devise an energy cycle using enthalpies of combustion of carbon, hydrogen and methane to calculate the enthalpy of formation.

The equations for the combustion of carbon, hydrogen and methane are shown below (the enthalpies of combustion are from the SQA data booklet):

$$C(s) + O_2(g) \rightarrow CO_2(g) \qquad \Delta H_1 = -394 \text{ kJ mol}^{-1}$$

$$H_2(g) + \frac{1}{2}O_2(g) \rightarrow H_2O(\ell) \qquad \Delta H_2 = -286 \text{ kJ mol}^{-1}$$

$$CH_4(g) + 2O_2(g) \rightarrow CO_2(g) + 2H_2O(\ell) \qquad \Delta H_3 = -891 \text{ kJ mol}^{-1}$$

The equations can be arranged in the following energy cycle:

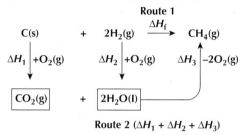

Route 2 ($\Delta H_1 + \Delta H_2 + \Delta H_3$)

Important notes on the energy cycle:

1. ΔH_2 has to be **multiplied by 2**, i.e. $2\Delta H_2$, because in the target equation 2 moles of hydrogen react.

2. ΔH_3 has to be **reversed**, i.e. $-\Delta H_3$, because in the cycle the carbon dioxide and water combine to form methane and oxygen, the reverse of combustion.

3. Applying Hess's law:

$$\Delta H_f = \Delta H_1 + 2\Delta H_2 - \Delta H_3$$
$$= -394 + (2 \times -286) - (-891)$$
$$\mathbf{\Delta H_f = -75 \text{ kJ mol}^{-1}}$$

TOP TIP

It is not always necessary to draw out an energy cycle. By writing out the equation for route 1 (target equation) and the equations for the enthalpy of reaction for each of the steps in route 2, rearranging them as required, and then applying Hess's law, the enthalpy of formation can be calculated.

Quick Test 41

1. The equation for the formation of ethene is $2C(s) + 2H_2(g) \rightarrow C_2H_4(g)$.

   Calculate the enthalpy of formation of ethene, using the enthalpies of combustion shown below:

$$C(s) + O_2(g) \rightarrow CO_2(g) \qquad \Delta H_1 = -394 \text{ kJ mol}^{-1}$$

$$H_2(g) + \frac{1}{2}O_2(g) \rightarrow H_2O(\ell) \qquad \Delta H_2 = -286 \text{ kJ mol}^{-1}$$

$$C_2H_4(g) + 3O_2(g) \rightarrow 2CO_2(g) + 2H_2O(\ell) \qquad \Delta H_3 = -1411 \text{ kJ mol}^{-1}$$

2. Calculate the enthalpy of formation of methanol from the enthalpy of combustion of carbon, hydrogen and methanol (use the SQA data booklet for enthalpies of combustion):

$$C(s) + 2H_2(g) + \frac{1}{2}O_2 \rightarrow CH_3OH(\ell) \ \Delta H_f \text{ (target equation)}.$$

Chemical energy 4

Bond and mean bond enthalpies

Bond enthalpy is the energy required to break one mole of bonds between the atoms in a mole of gaseous diatomic molecules, at standard temperature and pressure (25°C and 1 atmosphere).

For diatomic molecules such as hydrogen (H_2) and hydrogen chloride (HCl) there is only one possible bond energy for each, as the H–H bond and H–Cl bond can only exist in these molecules.

This is not the case for the likes of the C–C bond, which can exist in alkanes, cycloalkanes, alcohols etc. The C–C bond will have slightly different bond enthalpy depending on the type of molecule it exists in. The bond enthalpies quoted in these cases is the **mean bond enthalpy**, which is an average value.

Bond-breaking is an endothermic process, because energy has to be put in to overcome the force of attraction between the atoms.

Bond-making is an exothermic process – if energy needs to be put in to break a bond, then the same amount of energy should be given out when the same bond forms.

Estimating enthalpy changes

Bond enthalpies and mean bond enthalpies can be used to estimate the enthalpy changes for different reactions. The enthalpy of reaction is the difference between the energy needed to break bonds, and the energy released when bonds are made. When a fuel burns for example, the energy produced when bonds are made is greater than the energy needed to break bonds between reactant molecules.

Example

Burning hydrogen in oxygen:

Bond breaking	Bond making
H–H(g) + $\frac{1}{2}$O=O(g) $\rightarrow$	H–O–H(g)
436 + $\frac{1}{2}$(498)	2 (–463)
$\Delta H = 436 + 249$ +	(–926)
$\Delta H = -241 \cdot 0$ kJ mol^{-1}	

Example

Use bond enthalpies and mean bond enthalpies to estimate the energy released when 1 mole of methanol is burned.

$$CH_3OH(g) + \frac{3}{2}O_2(g) \rightarrow CO_2(g) + 2H_2O(g)$$

Worked answer:

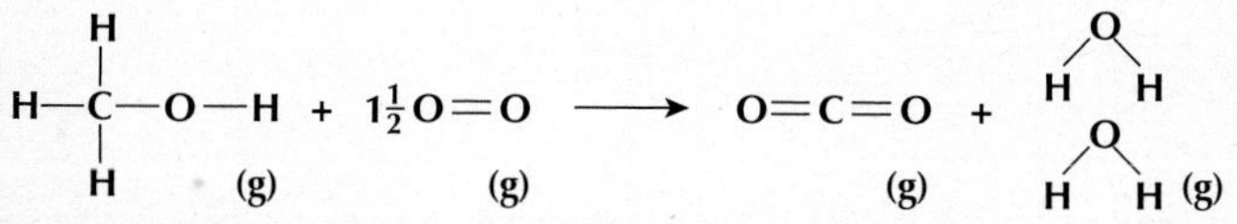

Total endothermic change for bond-breaking:

$3 \times$ C–H

$1 \times$ C–O

$1 \times$ O–H

$1\frac{1}{2} \times$ O=O

$= (3 \times 412) + 360$

$\quad + 463 + (1\frac{1}{2} \times 498)$

$= +2806$ kJ

Total exothermic change for bond-making:

$2 \times$ C=O

$4 \times$ O–H

$= (2 \times 743) + (4 \times -463)$

$= -3338$ kJ

Bond enthalpies in kJ mol^{-1}

C–H = 412

C–O = 360

O–H = 463

O=O = 498

C=O = 743

$$\Delta H_c = + 2806 - 3338 = -532 \text{ kJ mol}^{-1}$$

TOP TIP

Using this method gives an enthalpy of combustion which is regarded as an estimate and is different from the data booklet value. This is because not all of the reactants and products are gases in their standard states, and also the mean bond enthalpy values used are average values.

Quick Test 42

1. Use the bond enthalpies and mean bond enthalpies in the SQA data booklet to estimate the enthalpy of combustion of propane:

$$C_3H_8(g) + 5O_2(g) \rightarrow 3CO_2(g) + 4H_2O(g)$$

2. Use the bond enthalpies and mean bond enthalpies in the SQA data booklet to estimate the enthalpy of combustion of ethanol:

$$C_2H_5OH(g) + 3O_2(g) \rightarrow 2CO_2(g) + 3H_2O(g)$$

Oxidising and reducing agents 1

Oxidation and reduction

Many chemical reactions involve electrons being lost by one 'particle' and gained by another. The 'particle' can be an atom, molecule or ion. These types of reaction are known as redox reactions. The particle losing electrons is said to be oxidised. The particle gaining electrons is said to be reduced. Ion–electron equations can be written for oxidation and reduction, and they can be added to form the redox reaction.

TOP TIP

Memory aid for what happens to the electrons:

Oxidation   **R**eduction
Is   **I**s
Loss   **G**ain

Elements as oxidising and reducing agents

Metals tend to act as reducing agents because they lose electrons easily.

Example

In the thermite reaction, iron(III) ions are reduced to iron atoms and aluminium atoms are oxidised to aluminum(III) ions. The liquid iron can be run off to repair cracks in things like railway lines.

The iron(III) ions are reduced by gaining electrons from the aluminium atoms. The aluminium atoms are acting as **reducing agents**. When the aluminium atoms are oxidised and lose electrons to the iron(III) ions, the iron(III) ions are acting as **oxidising agents**.

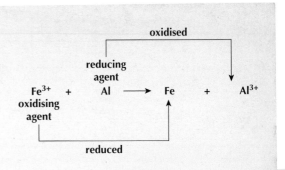

TOP TIP

Reducing agents lose electrons in a reaction and are themselves oxidised. Oxidising agents gain electrons in a reaction and are themselves reduced.

Example

When green chlorine gas is bubbled through colourless potassium bromide solution, the solution turns orange. The bromide ions are displaced from solution and form bromine molecules. The chlorine molecules are reduced to chloride ions.

The chlorine molecules are acting as **oxidising agents** and are themselves reduced to chloride ions. The bromide ions are acting as **reducing agents** and are themselves oxidised to bromine molecules.

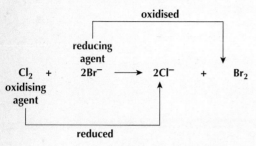

> **TOP TIP**
>
> Non-metal elements tend to act as oxidising agents because they accept electrons easily.

Reducing or oxidising agent?

Electronegativity values (see page 24) can be used to estimate whether an element is likely to be an oxidising or a reducing agent. Electronegativity is a measure of the attraction the atoms of an element have for electrons. The higher the value the greater the attraction for electrons. Fluorine, with a value of 4.0, will be a strong oxidising agent. The group 7 elements (halogens) are all strong oxidising agents.

The lower the electronegativity value, the more likely the element is to be a reducing agent. The group 1 metals (alkali metals) have low electronegativity values and are strong reducing agents.

> **TOP TIP**
>
> Electronegativity values are found in the SQA data booklet.

Quick Test 43

1. Identify the oxidising and reducing agents in the following equations:

   (a) $Zn(s) + Cu^{2+}(aq) \rightarrow Zn^{2+}(aq) + Cu(s)$

   (b) $Br_2(aq) + 2I^-(aq) \rightarrow 2Br^-(aq) + I_2(aq)$

   (c) $Cr(s) + NiSO_4(aq) \rightarrow CrSO_4(aq) + Ni(s)$

Oxidising and reducing agents 2

Compounds as oxidising and reducing agents

Carbon monoxide (CO) is an example of a compound which can act as a **reducing agent**.

Iron is extracted from its ore in a blast furnace by reaction with carbon monoxide. Iron (III) ions are reduced to iron:

$$Fe_2O_3(s) + 3CO(g) \rightarrow 2Fe(\ell) + 3CO_2(g)$$

The compound **hydrogen peroxide** (H_2O_2) is one of the most powerful **oxidising agents**.

In acidic conditions, iron(II) ions are oxidised to iron(III) ions by hydrogen peroxide, which is itself reduced. Ion–electron equations can be written for the oxidation and reduction reactions, which can then be added to give the overall redox equation.

$$2Fe^{2+}(aq) \rightarrow 2Fe^{3+}(aq) + 2e^- \text{ oxidation}$$

$$\underline{H_2O_2(\ell) + 2H^+(aq) + 2e^- \rightarrow 2H_2O(\ell) \text{ reduction}}$$

$$2Fe^{2+}(aq) + H_2O_2(\ell) + 2H^+(aq) \rightarrow 2Fe^{3+}(aq) + 2H_2O(\ell) \text{ redox}$$

Working out ion–electron equations

The reduction equation for hydrogen peroxide is more complex than previous examples encountered. These more complicated ion–electron equations involve hydrogen ions and water molecules, and can be worked out as follows:

1. Write down the main reactant(s) and product(s) for the reduction/oxidation. Make sure there are the same number of each element (not oxygen) on each side of the equation.

2. Add water to one side to balance the oxygen.

3. Add $H^+(aq)$ to the other side to balance the hydrogen atoms.

4. Add electrons to the same side as the $H^+(aq)$ ions, so that both sides of the equation have the same charge.

Applying these rules to hydrogen peroxide being reduced (acting as an oxidising agent):

1. $H_2O_2(\ell) \rightarrow H_2O(\ell)$
    There are two hydrogens on each side so there is no need to balance.

2. The oxygen atoms are not balanced. Add another water molecule to the right-hand side: $H_2O_2(\ell) \rightarrow 2H_2O(\ell)$

3. Add two $H^+(aq)$ ions to the left-hand side to balance the hydrogen atoms: $H_2O_2(\ell) + 2H^+(aq) \rightarrow 2H_2O(\ell)$

4. Add two electrons to the left-hand side to balance the 2+ charge so that there is zero charge on both sides of the equation: $H_2O_2(\ell) + 2H^+(aq) + 2e^- \rightarrow 2H_2O(\ell)$

TOP TIP

Although you have to know how to write ion–electron equations, many of them are listed in the electrochemical series in the SQA data booklet.

Group ions as oxidising and reducing agents

The **permanganate ion** (MnO_4^-) and **dichromate ion** ($Cr_2O_7^{2-}$) are examples of group ions, which act as strong **oxidising agents** when in an acidic solution.

Example

Acidified potassium permanganate (potassium manganate(VII)) solution reacting with iron(II) sulfate

The iron(II) ions are oxidised to iron(III) ions.

$Fe^{2+}(aq) \rightarrow Fe^{3+}(aq) + e^-$ oxidation

The permanganate ions act as oxidising agents and are reduced to manganese (II) ions.

$MnO_4^-(aq) \rightarrow Mn^{2+}(aq)$ reduction

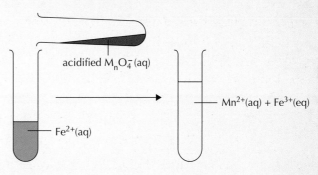

acidified $M_nO_4^-$ (aq)

Fe^{2+}(aq)

Mn^{2+}(aq) + Fe^{3+}(eq)

The ion–electron equation for the reduction reaction can be worked out using the balancing rules:

1. There is one Mn ion on each side so there is no need to balance.

2. The oxygens are not balanced so four H_2O need to be added to the right-hand side:
    $MnO_4^-(aq) \rightarrow Mn^{2+}(aq) + \textbf{4H}_\textbf{2}\textbf{O(ℓ)}$

3. Add eight $H^+(aq)$ to the left-hand side to balance the hydrogen atoms:
    $MnO_4^-(aq) + \textbf{8H}^\textbf{+}\textbf{(aq)} \rightarrow Mn^{2+}(aq) + \textbf{4H}_\textbf{2}\textbf{O(ℓ)}$

4. There is a total of 7+ charge on the left-hand side and 2+ on the right-hand side. Add five electrons to the left-hand side to balance the 2+ charge on the right:
    $MnO_4^-(aq) + \textbf{8H}^\textbf{+}\textbf{(aq)} + \textbf{5e}^\textbf{-} \rightarrow Mn^{2+}(aq) + \textbf{4H}_\textbf{2}\textbf{O(ℓ)}$

The two ion–electron equations can be combined to give the redox equation. The number of electrons in each ion–electron equation must be the same, so the oxidation equation must be multiplied by five:

$$5Fe^{2+}(aq) \rightarrow 5Fe^{3+}(aq) + 5e^- \text{ oxidation}$$
$$\underline{MnO_4^-(aq) + 8H^+(aq) + 5e^- \rightarrow Mn^{2+}(aq) + 4H_2O(ℓ) \text{ reduction}}$$
$$MnO_4^-(aq) + 8H^+(aq) + 5Fe^{2+}(aq) \rightarrow Mn^{2+}(aq) + 5Fe^{3+}(aq) + 4H_2O(ℓ) \text{ redox}$$

Note: The potassium and sulfate ions do not take part in the reaction – they are spectator ions.

Example

Acidified potassium dichromate reacting with tin(II) chloride

The tin(II) ions are reduced to tin(IV) ions.

$$Sn^{2+}(aq) \rightarrow Sn^{4+}(aq) + 2e^- \quad \text{oxidation}$$

The dichromate ions act as oxidising agents and are reduced to chromium(III) ions.

$$Cr_2O_7^{2-}(aq) + 14H^+(aq) + 6e^- \rightarrow 2Cr^{3+}(aq) + 7H_2O(\ell) \quad \text{reduction}$$

The two ion–electron equations can be combined to give the redox equation. The number of electrons in each ion–electron equation must be the same, so the oxidation equation must be multiplied by three:

$$3Sn^{2+}(aq) \rightarrow 3Sn^{4+}(aq) + 6e^- \quad \text{oxidation}$$

$$\underline{Cr_2O_7^{2-}(aq) + 14H^+(aq) + 6e^- \rightarrow 2Cr^{3+}(aq) + 7H_2O(\ell) \quad \text{reduction}}$$

$$Cr_2O_7^{2-}(aq) + 14H^+(aq) + 3Sn^{2+}(aq) \rightarrow 3Sn^{4+}(aq) + 2Cr^{3+}(aq) + 7H_2O(\ell) \quad \text{redox}$$

TOP TIP

The reduction ion–electron equations in the previous examples show the need for the oxidising agents to be acidified.

Example

Iodine in solution can be reduced by the sulfite ion:

TOP TIP

The **sulfite ion** (SO_3^{2-}) ion is an example of a group ion that can act as a **reducing agent**.

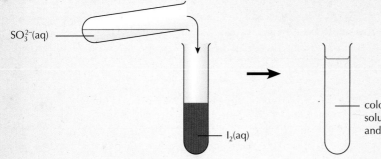

$SO_3^{2-}(aq)$

$I_2(aq)$

colourless solution containing $I^-(aq)$ and $SO_4^{2-}(aq)$

$$I_2(aq) + 2e^- \rightarrow 2I^-(aq) \text{ reduction}$$
$$\underline{SO_3^{2-}(aq) + H_2O(\ell) \rightarrow SO_4^{2-}(aq) + 2H^+(aq) + 2e^- \text{ oxidation}}$$
$$SO_3^{2-}(aq) + H_2O(\ell) + I_2(aq) \rightarrow SO_4^{2-}(aq) + 2H^+(aq) + 2I^-(aq) \text{ redox}$$

Note here that the number of electrons in each equation is the same, so the equations don't need to be multiplied up.

Although the oxidation equation for the oxidation of the sulfite ion can be found in the electrochemical series in the SQA data booklet, it can also be worked out by applying the four rules used previously:

1. $SO_3^{2-}(aq) \rightarrow SO_4^{2-}(aq)$

   There is one sulfur on each side, so there is no need to balance.

2. The oxygens are not balanced, so add one water to the left-hand side:

   $SO_3^{2-}(aq) + H_2O(\ell) \rightarrow SO_4^{2-}(aq)$

3. Add two $H^+(aq)$ to the right-hand side to balance the hydrogen atoms:

   $SO_3^{2-}(aq) + H_2O(\ell) \rightarrow SO_4^{2-}(aq) + 2H^+(aq)$

4. Add two electrons to the right-hand side to balance the 2+ charge:

   $SO_3^{2-}(aq) + H_2O(\ell) \rightarrow SO_4^{2-}(aq) + 2H^+(aq) + 2e^-$

Quick Test 44

1. Chlorine reacts with aqueous thiosulfate ions ($S_2O_3^{2-}$) which form sulfate ions (SO_4^{2-}). The unbalanced equation is shown:

   $$Cl_2(aq) + S_2O_3^{2-}(aq) \rightarrow Cl^-(aq) + SO_4^{2-}(aq)$$

   (a) Write the ion–electron equation for the reduction reaction.

   (b) Write the ion–electron equation for the oxidation reaction.

   (c) Combine the oxidation and reduction equations to give the balanced redox equation.

   (d) Identify the oxidising and reducing agents in the reaction.

2. In each of the following examples complete the ion–electron equations and state whether it is oxidation or reduction.

   (a) $S_4O_6^{2-}(aq) \rightarrow S_2O_3^{2-}(aq)$

   (b) $V^{3+}(aq) \rightarrow VO_3^-(aq)$

Oxidising and reducing agents 3

The electrochemical series

The electrochemical series lists various substances in order of how well they act as reducing/oxidising agents.

Part of the electrochemical series		
	Reaction	
	$Li^+(aq) + e^-$	$\rightleftharpoons Li(s)$
	$Cs^+(aq) + e^-$	$\rightleftharpoons Cs(s)$
	$Rb^+(aq) + e^-$	$\rightleftharpoons Rb(s)$
	$K^+(aq) + e^-$	$\rightleftharpoons K(s)$
	$Ca^{2+}(aq) + 2e^-$	$\rightleftharpoons Ca(s)$
	$Na^+(aq) + e^-$	$\rightleftharpoons Na(s)$
	$Br_2(\ell) + 2e^-$	$\rightleftharpoons 2Br^-(aq)$
	$O_2(g) + 4H^+(aq) + 4e^-$	$\rightleftharpoons 2H_2O(\ell)$
The strongest oxidising agents are at the bottom-left.	$Cr_2O_7^{2-}(aq) + 14H^+(aq) + 6e^-$	$\rightleftharpoons 2Cr^{3+}(aq) + 7H_2O(\ell)$
	$Cl_2(g) + 2e^-$	$\rightleftharpoons 2Cl^-(aq)$
	$MnO_4^-(aq) + 8H^+(aq) + 5e^-$	$\rightleftharpoons Mn^{2+}(aq) + 4H_2O(\ell)$
	$F_2(g) + 2e^-$	$\rightleftharpoons 2F^-(aq)$

The strongest reducing agents are at the top-right.

Using the electrochemical series

The electrochemical series can be used to predict whether a redox reaction will occur. Just looking at the substances involved alone gives no indication as to whether they will react.

The general rule is: if an oxidising agent is below the reducing agent in the electrochemical series, then a reaction will occur.

Example

Will potassium permanganate ($KMnO_4$) be able to oxidise bromide ions?

From the electrochemical series we see that the MnO_4^- is bottom-left, and therefore is a strong oxidising agent (easily reduced); Br^- ions are above on the right-hand side, so the reaction will occur.

The ion–electron equations are obtained from the electrochemical series.

The permanganate ion is reduced so the equation is written as it is in the electrochemical series:

$$MnO_4^-(aq) + 8H^+(aq) + 5e^- \rightarrow Mn^{2+}(aq) + 4H_2O(\ell)$$

The bromide ions are on the right-hand side so the equation must be reversed:

$$2Br^-(aq) \rightarrow Br_2(aq) + 2e^-$$

Note that the potassium ion (K^+) is top left in the electrochemical series so would not be able to act as a reducing agent – it is a spectator ion.

Using oxidising agents

Hydrogen peroxide is an extremely strong oxidising agent and acts as a very effective 'bleach' as it is able to break down coloured compounds. It is used to bleach wool, cotton and paper as well as lighten hair and even whiten teeth.

The strong oxidising power of **potassium permanganate** makes it useful in a number of areas, particularly as an antiseptic. Dilute solutions are used to treat mild skin conditions and fungal infections of the hands and feet, and can also cause some viruses to become inactive. Potassium permanganate is used in the water treatment industry to remove iron and hydrogen sulfide (rotten egg smell). It can be used in fish farms to kill parasites which infect the fish. Some florists use potassium permanganate for extending the life of fresh-cut flowers. It is also added for its disinfectant properties, as it can help to control algae spread in the water.

Quick Test 45

1. Look at the electrochemical series in the SQA data booklet and identify the strongest reducing agent and the strongest oxidising agent in the table.

2. (a) Use the electrochemical series in the SQA data booklet to predict whether the following reactions will take place:

    (i) $Br_2(aq) + 2I^-(aq) \rightarrow 2Br^-(aq) + I_2(aq)$

    (ii) $2Ag^+(aq) + Cu(s) \rightarrow 2Ag(s) + Cu^{2+}(aq)$

    (iii) $MnO_4^-(aq) + 2Cl^-(aq) \rightarrow Mn^{2+}(aq) + Cl_2(aq)$

    (iv) $F_2(g) + Cr_2O_7^{2-}(aq) \rightarrow 2Cr^{3+}(aq) + 2F^-(aq)$

    (b) For the reactions in (a) which do take place, use the electrochemical series to write oxidation and reduction ion–electron equations.

    (c) Combine the ion–electron equations in (b) to form redox equations.

    (d) Identify the oxidising and reducing agents in each reaction.

Chemical analysis 1

Chromatography

Chromatography is a technique which can be used to separate, identify and in some instances, obtain individual substances from complex mixtures. There are many different chromatographic methods including paper, thin layer, gas and gas-liquid.

How does chromatography work?

The principles for each type of chromatography are the same. The differences in **size** and **polarities** of molecules are used to separate them. A liquid or gas (the **mobile phase**) carries the molecules of a substance over or through a solid or a liquid absorbed onto a solid (the **stationary phase**).

Two methods commonly used in a school laboratory are **paper chromatography** and **thin layer chromatography** (TLC).

In paper chromatography, water on the surface of the paper acts as the stationary phase. In TLC, glass plates are coated with silica gel or alumina, which have very polar hydroxyl groups on their surface, and act as the stationary phase. Spots of the mixture are put onto the paper or plate, and they are placed in a suitable solvent – the mobile phase.

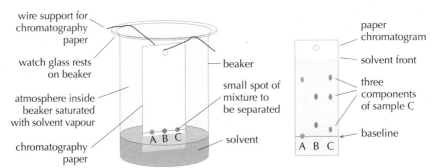

As the solvent begins to move up the paper or TLC plate, it dissolves the compounds that have been spotted on the chromatogram. The compounds will then move up the chromatogram as the solvent moves upwards. Molecules continually change from being in the mobile phase, i.e. dissolved in the solvent and moving, to being stopped due to intermolecular attractions with the molecules of the stationary phase.

TOP TIP

The speed at which the compounds move depends on two factors:
- how attracted they are to the stationary phase;
- their solubility in the mobile phase (the solvent).

Identifying what's in a mixture

Components of mixtures separated by paper or thin layer chromatography can sometimes be observed because they are coloured, e.g. the inks in felt-tip pens. More often they have to be developed in some way. The diagram on page 113 shows how two amino acids can be separated and then identified.

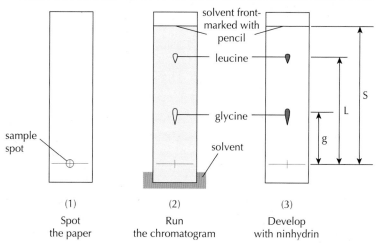

(1)   (2)   (3)

Spot   Run   Develop

the paper   the chromatogram   with ninhydrin

Separating and identifying the amino acids leucine and glycine in a sample.
Amino acids are colourless and become visible when sprayed with ninhydrin.

Components of a mixture can also be identified by the distance they travel compared to the distance travelled by the solvent – the **solvent front**. This is called the R_f **value** (the retention or retardation factor). The R_f value is constant for a particular substance if other conditions such as paper and solvent used remain the same.

Calculating the R_f value

$$R_f = \frac{\text{distance travelled by compound}}{\text{distance travelled by solvent}}$$

For the chromatogram above, the R_f value for leucine is:

$$R_f = \frac{L}{S}$$

$$= \frac{3 \cdot 5}{4 \cdot 5}$$

$$R_f = 0 \cdot 78$$

TOP TIP

R_f values have no units. Different compounds have different R_f values.

For molecules with similar polarity the larger the molecule, the smaller the R_f value, i.e. the slower it will move up the plate or paper.

A polar molecule will have a higher R_f value in polar solvents than it will have if a non-polar solvent is used.

Quick Test 46

1. Two compounds, A and B, are to be separated using paper chromatography. A is very soluble in the mobile phase and almost insoluble in the stationary phase. B has greater solubility in the stationary phase than in the mobile phase.

   (a) Which will travel further during chromatography?

   (b) Explain your answer to (a).

2. (a) Suggest why, in the chromatogram on page 112 one of the components in mixture A has not moved off the baseline.

   (b) Two of the mixtures in the chromatogram on page 112 contain one component which is the same. State which two and explain how you reached your conclusion.

3. (a) Calculate the R_f value for glycene in the chromatogram above if it travels 1·7 cm from the baseline.

   (b) Explain why R_f values can be used to identify a component in a mixture.

Chemical analysis 2

Using chromatography in analysis

Chromatography is used in laboratories to aid chemical analysis in many different situations.

Area	Use
Pharmaceutical companies	Determines amount of each chemical found in new product.
Hospitals	Detects chemicals in a patient's blood stream.
Law enforcement	Compares a sample found at a crime scene to samples from suspects.
Environmental agencies	Determines the level of pollutants in the water supply.
Manufacturing	Follows the progress of a reaction and purifies a chemical needed to make other products such as medicines.

Paper chromatography

One use of paper chromatography is to separate and identify amino acids in a mixture.

Sometimes the amino acids in a mixture are not separated completely because they have similar solubilities in a particular solvent. To overcome this problem the paper is rotated through 90°C and a different solvent is used to ensure a separation. This is known as two-way separation.

1. The mixture and known amino acids are spotted on the paper. The baseline is drawn using pencil, as ink is likely to separate out with the solvent.

2. As the solvent moves up the page the amino acids move with it and separate out. At this stage the amino acids are invisible.

3. When the solvent front reaches the top of the paper, the paper is removed, dried and sprayed with ninhydrin to show up the amino acids as purple spots.

4. If the separation is not good enough the experiment is repeated and a two-way chromatogram is run.

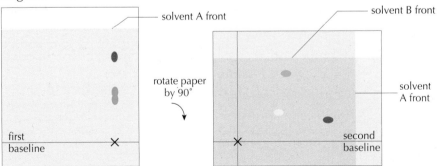

Two-way paper chromatography can be used to improve the separation of components of a mixture. The green components in the first chromatogram are separated out in the second and appear as yellow and blue (colours are used here to make the separation clearer to see).

Thin layer chromatography (TLC)

The process of separation is similar to paper chromatography. However, TLC has the advantage of being faster and gives better separation. There are also several choices of the stationary phase, such as alumina and silica, and it is relatively cheap. TLC can be used in forensic science to identify explosive residues and to detect cannabis. The separated components can be retrieved by scraping off the spots on the stationary phase and dissolving them in a suitable solvent.

Gas chromatography (GC) and gas-liquid chromatography (GLC)

Gas chromatography (GC) and gas-liquid chromatography (GLC) are techniques which can separate mixtures of gases and volatile liquids. These techniques can be used to separate very small quantities. GLC has become a standard method for detecting banned substances in sport, and for accurately determining the amount of alcohol in a motorist's blood when it is near the legal limit.

A gas-liquid chromatogram is usually a graph with peaks appearing at different times depending on the retention time of each component in a sample. The graphs can be compared to the graph obtained for known samples.

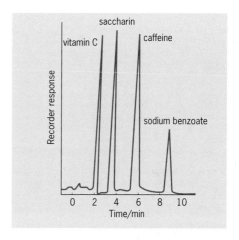

TOP TIP

The amount of substance present can be calculated from the area under each peak, not from the height of the peak. This is usually automatically calculated by a computer connected to the machine.

Quick Test 47

1. (a) Suggest why, when drawing the baseline on chromatography paper or a TLC plate, pencil is used.

   (b) Suggest why, when using two-way chromatography, two different mobile solvents are used.

2. (a) A database of retention times is kept for each GLC machine for known compounds. Suggest why it is essential to keep the conditions in the machine constant after forming the database.

   (b) By inspection of the chromatogram above, estimate which of the components there is most of and explain how you came to your conclusion.

Chemical analysis 3

Volumetric analysis

Volumetric analysis involves using a solution of accurately known concentration to determine the concentration of another substance. In order to do this, the concentration of the substance being used to react with the compound needs to be known accurately.

A solution of accurately known concentration is referred to as a **standard solution**.

Preparing a standard solution

Suppose you wish to prepare 500 cm³ of a 0·10 mol l⁻¹ sodium carbonate solution.

Step 1 Calculate the **mass** of sodium carbonate required.

$$\text{No of moles} = \text{concentration} \times \text{volume}$$
$$= 0·5 \times 0·1$$
$$= 0·05 \text{ mol}$$

$$\text{Mass required} = 0·05 \times 106 \text{ g}$$
$$= 5·30 \text{ g}$$

Step 2 Weigh out accurately the required mass and dissolve it in deionised water, in a small beaker.

Step 3 Pour the solution into a volumetric flask and rinse all of the solution out of the beaker using a water bottle.

Step 4 More deionised water is added until it reaches the etched line on the neck of the flask. The flask is stoppered and then inverted, to ensure the solution is fully mixed.

The concentration of the standard solution can be calculated if the exact mass weighed out is known – although the mass has to be measured accurately, it doesn't need to be exactly 5·30 g. Let's say 5·39 g was actually weighed out:

$$\text{moles} = \frac{\text{mass}}{\text{GFM}} = \frac{5·39}{106} = 0·051 \text{ mol}$$

$$c = \frac{n}{V} = \frac{0·051}{0·50} = 0·102 \text{ mol l}^{-1}$$

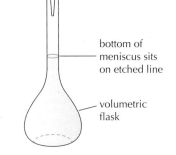

deionised water bottle

bottom of meniscus sits on etched line

volumetric flask

This standard solution is used in the calculation of the concentration of an acid in the example in **(1) Acid-base reactions** below.

Calculating an unknown concentration

(1) Acid-base reactions

The standard solution prepared above can be used in a titration, the results of

TOP TIP

You don't have to memorise the relationship between volume, concentration and balancing numbers – it is given in the SQA data booklet as: $\dfrac{c_1 V_1}{n_1} = \dfrac{c_2 V_2}{n_2}$

which can be used to calculate the concentration of another solution using the following relationship:

$$\frac{(\text{concentration} \times \text{volume})_{\text{alkali}}}{\text{balancing no.}_{\text{alkali}}} = \frac{(\text{concentration} \times \text{volume})_{\text{acid}}}{\text{balancing no.}_{\text{acid}}}$$

Example

19·7cm³ of 0·102 mol sodium carbonate solution was required to neutralise 25·0 cm³ of dilute hydrochloric acid solution. Find the concentration of the hydrochloric acid.

$$Na_2CO_3(aq) + 2HCl(aq) \rightarrow 2NaCl(aq) + CO_2(g) + H_2O(\ell)$$

Worked answer:

The balancing number for sodium carbonate is 1 and for hydrochloric acid is 2.

$$\frac{(\text{volume} \times \text{concentration})_{\text{acid}}}{\text{balancing no.}_{\text{acid}}} = \frac{(\text{volume} \times \text{concentration})_{\text{alkali}}}{\text{balancing no.}_{\text{alkali}}}$$

$$\frac{25 \cdot 0 \times \text{conc}}{2} = \frac{19.7 \times 0.102}{1}$$

$$12 \cdot 5 \times \text{conc} = 2 \cdot 01$$

$$\text{conc} = \frac{2.58}{12.5}$$

$$\text{conc} = 0 \cdot 161 \text{ mol l}^{-1}$$

TOP TIP

You don't have to change volumes to litres because volume appears on both sides of the relationship.

Redox titration

Volumetric analysis can be applied to a redox reaction. For example, the concentration of a solution of a reducing agent can be found using an oxidising agent.

Example

The concentration of iron(II) ions (reducing agent) in a solution can be found by titrating 20 cm³ portions of the solution with acidified potassium permanganate solution (oxidising agent) with a concentration of 0·101 mol l⁻¹.

The average titre was 13·7 cm³.

$$\text{Oxidation} \quad Fe^{2+} \rightarrow Fe^{3+}(aq) + e^-$$

Reduction $\quad MnO_4^-(aq) + 8H^+(aq) + 5e^- \rightarrow Mn^{2+}(aq) + 4H_2O(\ell)$

Redox $\quad MnO_4^-(aq) + 8H^+(aq) + 5 Fe^{2+}(aq) \rightarrow Mn^{2+}(aq) + 8H_2O(\ell) + 5 Fe^{3+}(aq)$

Worked answer:

$$\frac{(\text{concentration} \times \text{volume})_{\text{reducing agent}}}{\text{balancing no.}_{\text{reducing agent}}} = \frac{(\text{concentration} \times \text{volume})_{\text{oxidising agent}}}{\text{balancing no.}_{\text{oxidising agent}}}$$

$$\frac{20 \cdot 0 \times \text{conc Fe}^{2+}}{5} = \frac{13 \cdot 7 \times 0 \cdot 101}{1}$$

$4 \times \text{conc Fe}^{2+} = 1 \cdot 384$

$\text{conc Fe}^{2+} = \dfrac{1.384}{4}$

$\text{conc Fe}^{2+} = 0 \cdot 346 \text{ mol l}^{-1}$

TOP TIP

The balancing numbers are obtained from the redox equation. The oxidation equation has to be multiplied by 5 to balance the number of electrons in the reduction and oxidation equations.

In the reaction in the worked example the acidified permanganate acts as its own indicator. The point where the reaction has just reached completion is known as the **end point**. In a titration using potassium permanganate, the end point is indicated when a permanent pink colour is observed in the conical flask.

TOP TIP

Acidified potassium dichromate (orange) can also be used a self-indicator for redox titrations.

Example

A vitamin C tablet was dissolved in some deionised water and made up to 250 cm³ in a standard flask. 25 cm³ of the solution was titrated with 0·029 mol⁻¹ iodine solution. The average titre was 17·5 cm³.

Calculate the mass of vitamin C ($C_6H_8O_6$) in the tablet.

$$C_6H_8O_6 + I_2 \quad \rightarrow \quad 2I^- \quad + \quad C_6H_6O_6 \quad + \quad 2H^+$$

Worked answer:

Step 1: From the balanced equation, 1 mole of vitamin C reacts with 1 mole of iodine.

Calculate the number of moles of iodine from the information given. The number of moles of vitamin C reacting will be the same.

Moles of iodine = $c \times V$

$= 0 \cdot 029 \times 0 \cdot 0175$

$= 0 \cdot 00051 \text{ mol } (5 \cdot 1 \times 10^{-4} \text{ mol})$

This is the number of moles in the 25 cm³ in the sample titrated.

The tablet was dissolved in 250 cm³ of water so the number of moles of vitamin C will be 10 times this value, i.e. 0·0051 moles (5·1 x 10⁻³ mol).

Step 2: Calculate the gram formula mass (GFM) of the vitamin C and multiply it by the number of moles (found in step 1) to get the mass of vitamin C.

GFM of vitamin C ($C_6H_8O_6$) = 176 g

Mass of vitamin C in tablet = 176 × 0·0051
= 0·898 g

Quick Test 48

1. (a) Calculate the mass of sodium hydroxide required to make 500 cm³ of a standard solution with a concentration of 0·01 mol l⁻¹.

    (b) Describe how you would prepare the standard solution in part (a).

2. A 25 cm³ solution of sulfuric acid was titrated with a standard sodium hydroxide solution with a concentration of 0·109 mol l⁻¹. The average titre was 17·8 cm³. Calculate the concentration of the sulfuric acid.

$$2NaOH(aq) + H_2SO_4(aq) \rightarrow Na_2SO_4(aq) + 2H_2O(\ell)$$

3. Calculate the concentration of a hydrogen peroxide solution (H_2O_2) if 25 cm³ of the solution reacts completely with 14·8 cm³ of acidified potassium permanganate solution with a concentration of 0·11 mol l⁻¹.

$$2MnO_4^-(aq) + 6H^+(aq) + 5H_2O_2(\ell) \rightarrow 2Mn^{2+}(aq) + 8H_2O(\ell) + 5O_2(g)$$

Learning checklist

Getting the most from reactants

In this section you have learned:

- The chemical industry makes a huge contribution to our quality of life.
- When making new products, industrial chemists have to consider the most economical way to make them in order to maximise profit.
- Key considerations when making a new product are:
  - availability, sustainability and cost of feedstocks;
  - product yield;
  - marketability of by-products;
  - recyclability of reactants;
  - energy requirements;
  - use of processes that reduce or eliminate the use and production of hazardous substances (green chemistry).
- The key points of green chemistry are:
  - waste prevention;
  - safer product design and products;
  - use of renewable feedstocks;
  - designing products that will biodegrade.
- The percentage yield is a way of measuring the efficiency of a chemical process:

$$\text{percentage yield (\%)} = \frac{\text{actual yield}}{\text{theoretical yield}} \times 100$$

- The atom economy is a way to measure the efficiency of a reaction:

$$\text{atom economy (\%)} = \frac{\text{mass of desired product}}{\text{total mass of reactants}} \times 100$$

- Percentage yield and atom economy have to be considered together when deciding the efficiency of a process.
- The mole ratio of atoms/ionic units in a compound can be worked out from a chemical formula.
- The mole ratio of reactants and products can be worked out from a balanced equation.
- Masses/moles reacting and produced can be calculated from mole ratios in balanced equations.
- If the cost of feedstocks and the percentage yield are known for a reaction, then the cost of the reactants can be calculated.
- Masses, moles, concentrations and volumes of solutions reacting and being produced can be calculated from mole ratios in balanced equations.
- Reactants in excess and the limiting reactant can be calculated using the mole ratios in balanced equations.
- One mole of any gas occupies the same volume when measured at the same temperature and pressure, and is known as molar volume (V_m).

- Molar volume is measured in litres mol^{-1}.
- The volume of a gas can be calculated from the mass/number of moles and vice versa, given the molar volume.
- Volumes of reactant and product gases can be worked out from mole ratios in the balanced equation.

Equilibria

In this section you have learned:

- Many reactions are reversible.
- Reversible reactions attain a state of dynamic equilibrium when the rates of the forward and backward reactions are equal.
- Dynamic equilibrium will only be attained in a closed system where no products can escape, or reactants be added.
- At equilibrium the concentration of reactants and products remains constant, but rarely equal.
- Increasing the concentrations of reactants shifts the equilibrium to the right (products).
- Increasing the pressure shifts the equilibrium in the direction which results in a decrease in the number of moles of gas reactant or product.
- Increasing the temperature shifts the equilibrium in the direction of the endothermic reaction (ΔH = positive).
- Catalysts do not shift the equilibrium position, but allow equilibrium to be reached quicker.
- In industry, reaction conditions are changed so that product yield is maximised at the lowest cost.
- In industrial reactions, a balance has to be reached between shifting the equilibrium position towards products, and reaching the equilibrium position as quickly as possible.

Chemical energy

In this section you have learned:

- It is essential that industrial chemists be able to calculate the amount of heat given out or taken in by each chemical reaction in a chemical process.
- Endothermic reactions require heat to be supplied, which incurs costs.
- The heat produced during exothermic reactions often needs to be removed to avoid the temperature rising too high.
- The heat produced during exothermic reactions is not wasted and can be used elsewhere in a process, e.g. to heat catalysts.
- The enthalpy of combustion is the enthalpy change when 1 mole of a substance burns completely in oxygen.
- Enthalpy of combustion is always exothermic.

- Enthalpy of reaction experiments can be carried out in the laboratory. Enthalpy changes are calculated using $E_h = c \times m \times \Delta T$, where m = mass of water (kg); c = specific heat capacity of water ($4\cdot18$ kJ kg$^{-1\circ}$ C^{-1}); ΔT = change in temperature.
- Enthalpy of reaction is measured in kJ mol^{-1}.
- Enthalpies of formation can be used to calculate enthalpies of other reactions.
- How to use Hess's law to calculate enthalpy changes.
- For a diatomic molecule, XY, the molar bond enthalpy is the energy required to break one mole of XY bonds.
- The mean molar bond enthalpy is the average bond enthalpy for bonds that exist in different molecular environments.
- The difference between the energy needed to break bonds in gaseous reactants and the energy released when new bonds are made gives an estimate of the enthalpy change for the reaction.

Oxidising and reducing agents

In this section you have learned:
- An oxidising agent is a substance which accepts electrons.
- A reducing agent is a substance which donates electrons.
- Oxidising agents are reduced.
- Reducing agents are oxidised.
- Elements can act as reducing and oxidising agents.
- Elements with low electronegativity values, like the group 1 metals, are strong reducing agents.
- Elements with high electronegativity values, like the group 7 elements, are strong oxidising agents.
- How to write reduction and oxidation ion–electron equations involving elements.
- Compounds and group ions can act as oxidising and reducing agents.
- How to write reduction and oxidation ion–electron equations involving compounds and group ions.
- Given a group ion, work out the balanced ion–electron equation.
- How to balance reduction and oxidation ion–electron equations and combine them to form the redox equation.
- Recognise oxidising and reducing agents in a redox equation.
- Hydrogen peroxide, acidified permanganate and acidified dichromate solutions are strong oxidising agents.
- The oxidising property of hydrogen peroxide makes it useful as a bleaching agent, to remove colour from materials.
- The oxidising property of the permanganate ion makes it effective in killing fungi and bacteria, and can also deactivate viruses.

- The electrochemical series lists substances in order of how well they act as reducing/oxidising agents.
- Strong reducing agents are found at the top right of the electrochemical series.
- Strong oxidising agents are found at the bottom left of the electrochemical series.
- Ion–electron equations in the electrochemical series are written as reductions, but can be reversed to obtain the oxidation equation.

Chemical analysis

In this section you have learned:

- There is a range of chromatography techniques including paper, thin layer, gas and gas-liquid chromatography.
- Size and polarity of molecules are important factors in separating substances by chromatography.
- R_f values can be calculated.
- Volumetric analysis is a quantitative analysis technique.
- A standard solution is a solution of accurately known concentration.
- Volumetric analysis uses an accurately known concentration of one substance to find the concentration of another.
- Volumetric analysis includes the use of acid-base titrations and redox titrations.
- The end point of a titration is the point where the reaction has just reached completion.
- An indicator is a substance that changes colour at the endpoint of a reaction.
- In redox reactions some substances, e.g. potassium permanganate and potassium dichromate, can act as self-indicators.

Researching chemistry 1

Skills of scientific inquiry

The purpose of the **Researching chemistry** unit is to help you develop the skills of scientific inquiry – skills that are necessary to enable you to carry out chemical research. These skills will help you undertake the assignment that is part of the course assessment required by the SQA.

TOP TIP

The skills of scientific enquiry include:
- **researching current literature** to help you understand the underlying chemistry of a topic;
- carrying out **practical investigative work** associated with the topic;
- **communicating your findings** clearly to others.

Sourcing information

No longer do we need to rely only on books for information. The internet has become the number one resource when seeking information on any subject. The world wide web holds a vast bank of instantly available information on every topic you can think of.

Text books and scientific journals are very useful sources of information. However, the internet has become an essential resource when seeking information on any subject. Following some simple rules will help ensure that the information you find when surfing the internet for information is relevant and reliable.

- Bookmark sites that you might want to return to for information.
- Check the URL suffix. Generally, government and academic sites give **reliable** information. Commercial sites and non-profit organisation sites may well give factually correct information but may be written for a particular purpose.

  - .ac.uk   UK academic institution
  - .edu   US academic institution
  - .gov.uk   UK government
  - .org.uk   UK non-profit organisation
  - .co.uk   UK company or individual
  - .com   multinational company

- Evaluate the information.

TOP TIP

It is important to consider whether or not the information you find is relevant and reliable. Is it supported by information on other websites? Does the information appear biased in any way?

Common laboratory apparatus you should be familiar with

In order to carry out the practical investigative work you will need to be familiar with, and know when and how to use, the pieces of chemical apparatus listed in the table.

beaker	evaporating basin	test tubes / boiling tubes
burette	filter funnel	thermometers
conical flask	measuring cylinder	volumetric flasks
delivery tubes	Liebig condenser	
dropper	pipette and safety filler	

Techniques you should be familiar with

Filtration

Filtration is used to separate a solid residue from a liquid or solution.

Filtration

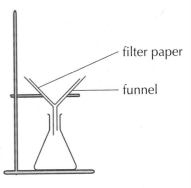

filter paper

funnel

Distillation

Distillation is used to separate a mixture of liquids – the liquid with the lower boiling point comes off first. A solvent can also be separated from a solution, collecting the solvent for reuse and leaving the solute in the distillation flask.

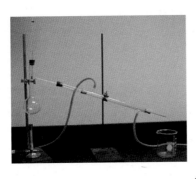

Distillation

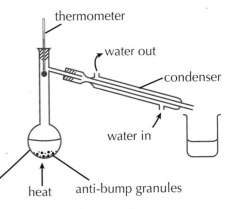

thermometer

water out

condenser

water in

distillation flask

heat

anti-bump granules

Using a balance

In most instances it is sufficient to use a **top-pan balance** and use the 'tare' facility when weighing out chemicals. The tare button will re-set the balance to zero, allowing the mass being weighed to be read directly from the display.

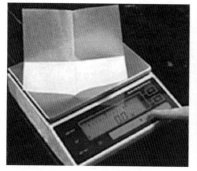

Collecting a gas

If the gas is insoluble in water then it can be collected by the **displacement of water**.

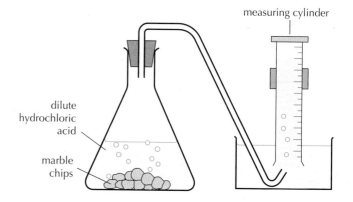

measuring cylinder

dilute hydrochloric acid

marble chips

If a gas is soluble in water or the, volume of gas needs to be measured accurately, then a gas syringe can be used.

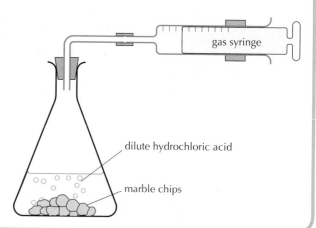

gas syringe

dilute hydrochloric acid

marble chips

1. Passing a gas through a liquid

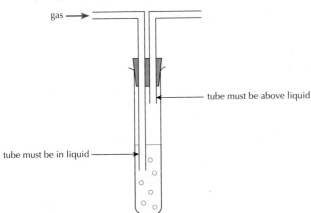

gas →

tube must be above liquid

tube must be in liquid

Often a gas is passed through a liquid as part of the separation process and through a solid, such as a drying agent, to remove water before it is collected. You may be asked to draw a diagram to show how both techniques could be carried out, and there are important points to note as highlighted in the diagrams.

2. Passing a gas through a drying agent

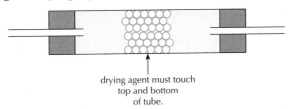

drying agent must touch
top and bottom
of tube.

Safe methods for heating

There are three common methods used in school chemistry laboratories for heating substances:

*   **Bunsen burner**;
*   **water bath**;
*   **heating mantle**.

Using a Bunsen burner will heat something rapidly; care needs to be taken, particularly if flammable liquids are involved.

It is often better to use a water bath or heating mantle when heating flammable liquids.

Quick Test 49

1. Barium sulfate can be produced by the following reaction:

$$BaCl_2(aq) + CuSO_4(aq) \rightarrow BaSO_4(s) + CuCl_2(aq)$$

Suggest the most suitable method for obtaining a sample of the barium sulfate.

2. Ammonia is a very soluble gas, which can be produced in the laboratory by heating an ammonium salt with a base. Describe how you would collect a sample of the gas in the laboratory.

Researching chemistry 2

Titration

To carry out a **titration** accurately it is important to know how to use a pipette and burette properly.

Using a pipette

A **pipette** is used to deliver a precise volume of liquid. The pipette should be rinsed with water and then a small volume of the liquid to be pipetted. The liquid is then drawn up into the pipette, to above the graduation mark on the stem. It is then removed from the liquid, and the liquid in the pipette is slowly released, until the bottom of the meniscus just touches the graduation mark.

> **TOP TIP**
>
> Although the word 'liquid' is often used in a description, it can also be used to mean 'solution'.

> **TOP TIP**
>
> A safety filler should always be used to fill a pipette as shown in the photo opposite.

Using a burette

A **burette** is used to gradually and accurately add small volumes of solution to a reaction vessel, normally a conical flask. The burette should be rinsed with the liquid to be used and then clamped vertically in a burette stand.

The burette is filled to above the scale and then the tap opened to allow the liquid or solution to run down onto the scale. You must be careful to ensure that no air bubbles are trapped in the jet of the burette. The bottom of the meniscus is used to read the initial volume from the scale. It is important that your eye level is the same as the liquid level.

> **TOP TIP**
>
> A filter funnel can be used when filling a burette, but this must be removed before taking the initial burette reading and carrying out the titration.

Carrying out a titration

The correct procedure for carrying out a titration is to swirl the titration flask at the same time as adding the liquid from the burette. The stopcock of the burette is operated using the left hand if you are right-handed, and the flask swirled with the right hand. This can be difficult and requires practice. A white tile is used to help identify the permanent colour change that indicates the endpoint of the titration.

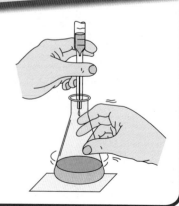

> **TOP TIP**
>
> As the endpoint is approached the liquid from the burette is added dropwise.

Analysis of data

Numerical results from experiments often need to be processed in some way. The data must be meaningful and correct. The table opposite gives some results from titrations carried out by a group of students.

Titration 1 is the rough titration and should not be used.

Titration 2 is a 'rogue' result. i.e. it is way out compared to the others, perhaps because the student has made a mistake with the volumes of the other reactant used.

Titrations 3 and 4 are concordant – they are within 0·1 cm³ of each other and can be used to calculate the average titre. These results are said to be reproducible.

Titration	Volume (cm³)
1	19·8
2	26·3
3	17·7
4	17·6

TOP TIP

You should be able to calculate averages, eliminate rogue data and draw lines/curves of best fit.

Reproducibility

For an experiment to be classed as reproducible it has to provide the same results, within experimental error, when repeated. The table opposite gives some results for a titration which are not reproducible because none of the results are close enough to each other.

Titration	Volume (cm³)
1	18·8
2	22·3
3	15·7
4	17·6

Quick Test 50

1. Calculate the average titre that should be used in a calculation from the results in the table at the top of the page.

2. The alcohol content of wine was analysed by four students. Each student carried out the experiment three times.

	Experiment 1 (%)	Experiment 2 (%)	Experiment 3 (%)
Student A	10·0	9·0	8·0
Student B	6·4	6·5	6·5
Student C	6·5	6·7	6·9
Student D	9·0	8·5	9·6

Which student obtained the most reproducible results?

Glossary

ΔH: see **enthalpy change**.

ΔH_c: see **standard enthalpy of combustion**.

ΔH_f: see **standard enthalpy of formation**.

E_a: see **activation energy**.

activated complex: a stage reached in a reaction where an intermediate product is formed.

activation energy (E_a): the minimum amount of kinetic energy needed by reactants before reaction can occur.

aldehydes: a flavour molecule with the carbonyl functional group on the end carbon.

amino acids: molecules with an amino group (**–NH$_2$**) and a carboxyl group (**–COOH**).

antioxidants: reduce the oxidation of chemicals in foods.

atom economy: compares the proportion of reactant atoms that end up in a useful product to the number that end up as waste:

$$\text{atom economy (\%)} = \frac{\text{mass of desired product}}{\text{total mass of reactants}} \times 100$$

bond enthalpy: the energy required to break one mole of bonds between the atoms in a mole of gaseous diatomic molecules, at standard temperature and pressure (25°C and 1 atmosphere).

bonding continuum: a scale that has ionic bonding at one end and pure covalent at the other, with polar covalent in between.

carbonyl group:

$$\begin{array}{c} O \\ \| \\ C \\ /\quad\backslash \end{array}$$

chromatography: a technique which can be used to separate, identify and in some instances, obtain individual substances from complex mixtures.

collision geometry: the angle at which molecules collide.

collision theory: simple collision theory states that for reactants to form products they must first come in contact with each other (collide).

covalent network: a giant 3D structure in which all of the atoms are covalently bonded to each other.

covalent radius: half the distance between the nuclei of two atoms joined by a single covalent bond.

denature: breaking of hydrogen bonds in protein resulting in loss of shape of the molecules.

diols: alcohols with two hydroxyl groups.

dynamic equilibrium: the point where, in a reversible reaction, the rate of the forward reaction equals the rate of the reverse reaction.

electronegativity: the attraction an atom involved in a bond has for the electrons of the bond.

emulsifier: substance which enables normally immiscible liquids to mix.

emulsion: a mixture of two or more liquids which are normally immiscible.

energy distribution diagram: a graph of the number of molecules against kinetic energy, which shows how the energy of the reactants varies at a particular temperature.

enthalpy change (ΔH): change in energy which accompanies a chemical reaction.

enzyme: biological catalyst; most enzymes are proteins.

essential amino acids: amino acids needed by the body but which the body cannot make so gets from food.

essential oils: the concentrated extracts of the volatile non-water soluble aroma compounds found in plants.

ester link: bond formed between a carboxylic acid molecule and alcohol molecule when they react to form an ester.

fats and oils: naturally occurring esters found in animals and plants.

fatty acids: long-chain carboxylic acid found in fats and oils.

fragrances: pleasant, sweet-smelling smells, caused by essential oils in plants.

free radicals: reactive particles which have unpaired electrons.

free radical chain reaction: reaction in which free radicals react in three distinct phases:
(1) **initiation**: the first step where free radicals are formed when a molecule absorbs radiation.
(2) **propagation**: steps where free radicals react to form further free radicals that can themselves react.
(3) **termination**: step in which free radicals combine, slowing the rate and stopping the reaction.

free radical scavengers: substances that remove free radicals and stop chain reactions.

fullerene: a molecular form of carbon (C_{60}).

glycerol: propane-1,2,3-triol, an alcohol with three hydroxyl groups, found in fats and oils.

green chemistry: the design of chemical products and processes that reduce or eliminate the use and production of hazardous substances.

hydrogen bond: when hydrogen is bonded to oxygen, nitrogen or fluorine it results in a very strong permanent dipole, which in turn results in a strong permanent dipole–permanent dipole attraction between molecules called a hydrogen bond.

hydrolysis: the breaking down of a compound which involves the addition of the elements in water.

hydrophobic: water-hating.

hydrophilic: water-loving.

initiation: see **free radical chain reaction**.

ionisation energy: the energy required to remove an electron from every atom in a mole of atoms in the gaseous state.

intermolecular forces: attractions between molecules.

intramolecular forces: attractions between atoms within a molecule.

ketones: flavour molecules where the carbonyl group is not on the end carbon.

London dispersion forces: forces of attraction caused by temporary dipoles in neighbouring atoms or molecules.

mean bond enthalpy: the average bond energy, taking into account the environment of the atoms forming the bond.

miscibility: the ability of liquids to mix in all proportions, forming a solution.

molar volume (V_m): the volume occupied by one mole of a gas when measured at a given temperature and pressure.

monatomic: a substance that exists as individual non-bonded atoms (the noble gases).

oxidising agent: accepts electrons from a reactant and so oxidises it.

peptide link: bond formed between amino acids when they react – also known as the amide link.

percentage yield: compares the expected product quantity with the actual amount produced:

$$\text{percentage yield} = \frac{\text{actual yield}}{\text{theoretical yield}} \times 100$$

permanent dipole–permanent dipole interactions: forces of attraction occurring between molecules with permanent dipoles.

polar covalent: the bond formed when two atoms share electrons unequally resulting in permanent dipoles.

primary alcohols: the carbon atom with the hydroxyl group is attached to **one** other carbon atom.

propagation: see **free radical reaction**.

proteins: natural condensation polymers made up of amino acids.

pure covalent: the bond formed when two atoms share bonding electrons equally.

R_f value: in chromatography, the distance travelled by the components of a mixture compared to the distance travelled by the solvent:

$$R_f = \frac{\text{distance travelled by compound}}{\text{distance travelled by solvent}}$$

reducing agent: supplies electrons to a reactant in order to reduce it.

reversible reaction: a reaction which can take place in both directions, i.e. the products can re-form reactants.

secondary alcohols: the carbon atom with the hydroxyl group is attached to **two** other carbon atoms.

soap: salt of long-chain fatty acids made from fats and oils.

solvent front: the distance travelled by the solvent in chromatography.

Glossary

standard enthalpy of combustion (ΔH_c): the enthalpy change when 1 mole of a substance burns completely, measured at standard temperature (25°C) and pressure (1 atmosphere), all substances being in their standard states.

standard enthalpy of formation (ΔH_f): the enthalpy change when one mole of a compound is formed from its elements in their standard states under standard conditions (25°C and 1 atmosphere).

sun block: mixtures containing compounds which reflect UV light and stop it reaching the skin.

termination: see **free radical reaction**.

terpenes: molecules that can be viewed as being based on isoprene, (2-methylbuta-1, 3-diene), units joined together.

tertiary alcohols: the carbon atom with the hydroxyl group is attached to **three** other carbon atoms.

triols: alcohols with three hydroxyl groups.

ultraviolet (UV) light: high-energy radiation which causes many chemical reactions in our skin.

van der Waals forces: the three types of intermolecular force: London dispersion forces, permanent dipole–permanent dipole interactions and hydrogen bonding.

Answers to Quick Tests

Quick Test 1

1. (a) The more particles there are, the more collisions there will be, and the greater the chance of reaction and products being formed.

   (b) Small particles offer a bigger surface area, so there are likely to be more collisions; therefore, the the chance of reaction and products being formed is greater.

2. The angle at which the reactants collide can affect the rate of reaction. When diatomic molecules collide side-on, the collision is likely to be more successful than colliding end-on.

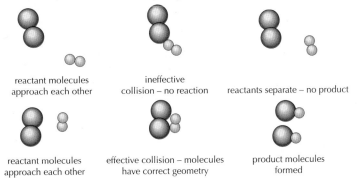

Quick Test 2

1. Sparks might be produced, which could provide the activation energy needed for the gas to react and there could be an explosion.

2. (a)

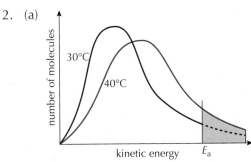

   (b) The shaded area represents the number of particles with energy equal to or greater than the activation energy for the reaction.

3. The rate of the reaction increases when the temperature is increased because more particles have a kinetic energy equal to or greater than the activation energy.

Quick Test 3

1. (a) $0.015 \ s^{-1}$

   (b) (i) Around $0.1 \ s^{-1}$

      (ii) For this reaction rate is directly proportional to concentration – doubling the concentration doubles the rate.

2. 40 s

3. $0.01 \ s^{-1}$

Quick Test 4

1. (a)

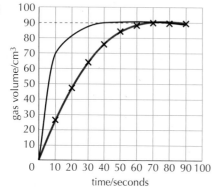

(b) Initially the reaction is very fast (the slope is steep), because there is a much larger surface area when powder is used. The larger the surface area, the more chance there is of successful collisions between reactants. As reactants are used up, the rate slows and the slope of the graph becomes less steep. Eventually the reaction stops because the acid is used up. The total volume of gas produced is the same as Experiment 1 because the same quantities are reacted – the final volume of gas is reached more quickly with powder.

Quick Test 5

1. (a) (i) −20 kJ   (ii) 15 kJ

   (b) (i) Exothermic

      (ii) The products have less PE than the reactants so energy is given out to the surroundings.

2. (a) (i) and (ii)

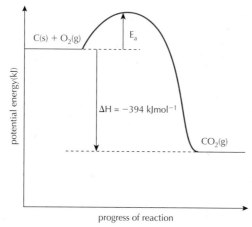

   (iii) Exothermic – indicated by − ΔH value.

(b) (i) and (ii)

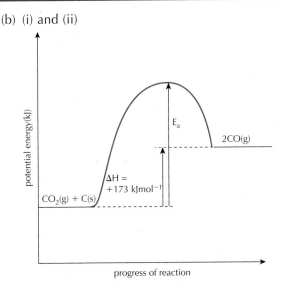

(iii) Endothermic – indicated by $+\Delta H$ value.

Quick Test 6

1. (a) The activated complex is an intermediate product formed during a chemical reaction. In a PE diagram the top of the energy barrier represents the point at which the activated complex is formed.

   (b) A----A
       ⋮   ⋮
       B----B

2. (a) Adsorption

   (b) As reactants pass over a catalyst at least one temporarily attaches to the surface of the catalyst (adsorption). The bonds in the reactant molecules weaken and new bonds form at the surface of the catalyst. Product molecules finally leave the surface of the catalyst (desorb).

   (c) A catalyst offers a reaction pathway in which each of the steps has a lower activation energy than the uncatalysed reaction, so more molecules have energy equal to or higher than the activation energy of the catalysed pathway.

Quick Test 7

1. These forces of attraction acting between atoms of noble gases are known as London dispersion forces. They are caused by the continual movement of electrons in an atom, which causes a temporarily uneven distribution of charge at opposite sides of an atom – known as a temporary dipole. This means that one side of the atom is temporarily slightly negative (δ^-), which results in the other side being temporarily slightly positive (δ^+). This in turn induces (causes) a temporary dipole in a neighbouring atom. This results in the δ^- side of one atom attracting the δ^+ side of a neighbouring atom, so a force of attraction is formed between them.

2. (a) They both increase.

   (b) Halogens are diatomic molecules which are held together by London dispersion forces. As you go down the group the molecules have more electrons so the strength of the London dispersion forces increases. This means more energy is needed to separate the molecules so the melting and boiling points increase.

Quick Test 8

1. Sulfur consists of S_8 molecules and phosphorus P_4. The sulfur molecules have more electrons than the phosphorus molecules and so have stronger London dispersion forces. This in turn means they have a higher melting point than phosphorus.

2. Fullerenes have molecules with 60 carbons or more arranged in ball shapes. The carbons are held together by covalent bonds (intramolecular forces) and the molecules are held together by London dispersion forces. Because the molecules are so big they have a large number of electrons. This results in the London dispersion forces being very strong, so a lot of energy is needed to separate the molecules, so their sublimation point is high.

3. (a) In diamond each carbon atom is covalently bonded to four other carbon atoms in a tetrahedral arrangement. All the outer electrons in the atom of each carbon are used to make single covalent bonds with neighbouring atoms, which results in a giant covalent network. In order to separate the atoms in a covalent network strong covalent bonds have to be broken, which requires a lot of energy so the sublimation point is very high.

   Sulfur is molecular and the molecules are held together by weak London dispersion forces so it does not require a lot of energy to separate the molecules so the boiling point is low compared to diamond.

   (b) In graphite each carbon atom is bonded to only three other carbon atoms. The fourth electron in each carbon atom is delocalised, so graphite can conduct electricity.

   (c) Silicon, like carbon, is in group 4 of the periodic table. Silicon forms four covalent bonds with four other silicon atoms and forms a covalent network similar to diamond.

Quick Test 9

1. (i) group (ii) radius (iii) increases (iv) nucleus (v) decreases (vi) period (vii) covalent (viii) shielding (ix) stronger.

2. (a) First ionisation energy: $Mg(g) \rightarrow Mg^+(g) + e^-$

   Second ionisation energy: $Mg^+(g) \rightarrow Mg^{2+}(g) + e^-$

   (b) As electrons are being removed the number of protons in the nucleus remains the same, so the pull on the remaining electrons is increased, so it takes more energy to remove a second electron than the first.

   (c) To remove a third electron requires much more energy as it has to be removed from an energy level closer to the nucleus, which is held more tightly than the outer electrons.

   (d) $1451 + 738 = 2189$ kJ mol^{-1}

3. (i) down (ii) decreases (iii) radius (iv) electrons (v) nucleus (vi) across (vii) covalent (viii) shielding (ix) stronger.

Quick Test 10

1. (i) ionic – electronegativity difference greater than 2 (2·3)

   (ii) polar covalent – electronegativity difference less than 2 (1·0)

   (iii) polar covalent – electronegativity difference less than 2 (0·4)

2. When two atoms share bonding electrons equally it is known as pure covalent bonding. This is because they have the same attraction for the bonding electrons – both atoms have the same electronegativity value.

Atoms of elements with different electronegativity values form polar covalent bonds. This means that one atom has a greater attraction for the bonding electrons than the other. The atom with the greater attraction for the bonding electrons will have a slightly negative charge (δ^-), leaving the other atom with a slightly positive charge (δ^+). A permanent dipole is formed.

3. The type of bonding changes gradually as the difference in electronegativity between atoms changes. There is no clear distinction between covalent and ionic bonding.

Quick Test 11

1. Both have the same strength of London dispersion forces between molecules, but in addition ICl is polar covalent and so has a permanent dipole, which results in permanent dipole–permanent dipole interactions between molecules. This means more energy is needed to separate ICl molecules, so its melting point is higher.

2. (a) (i) Polar covalent.

    (ii) Polar.

    (iii) The shape of the molecule means that one side of the molecule is δ^+ and the other is δ^-. The molecule therefore has a permanent dipole.

    (b) Permanent dipole–permanent dipole interactions.

Quick Test 12

1. (a) (ii), (iii) and (iv)

    (b) All three have hydrogen attached to the atom of a very electronegative element, which gives rise to a very polar bond. The molecules of these compounds are held together by the attraction of $H^{\delta+}$ of one molecule for the $O^{\delta-}$ or $N^{\delta-}$ or $F^{\delta-}$ of another molecule.

2. (a) (i) Around $-80°C$

    (ii) Water molecules have hydrogen bonding between the molecules but the other molecules don't. More energy is needed to separate water molecules because of the hydrogen bonding.

    (b) When water it is cooled and reaches $4°C$ it begins to expand. When the water freezes and expands, the force causes to pipes to crack. This expansion of water is due to the ordering of the molecules into an open structure, with an increased number of hydrogen bonds.

3. (a) Potassium chloride exists as an ionic lattice. The water molecules interact with the ions in the lattice, break the electrostatic attraction between the ions and the ions go into solution. An electrostatic force of attraction is created between the ions and the polar water molecules – the ions are hydrated with water molecules.

    (b) Candle wax is a mixture of non-polar molecules and so water molecules can't interact with them. Heptane is a non-polar solvent. London dispersion forces can form between the molecules.

Quick Test 13

1. (a) (i) propyl methanoate

   (ii) ethyl butanoate

   (b) (i)

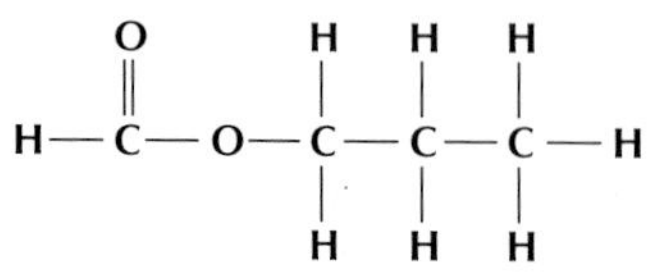

   (ii)

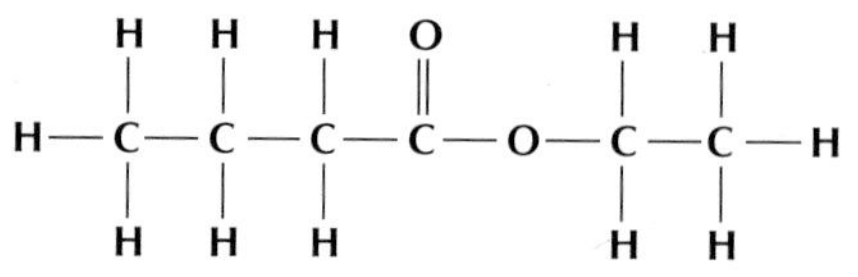

2. (a)

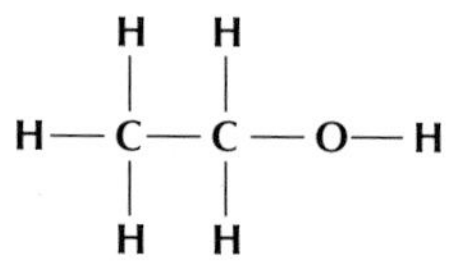

   ethanol 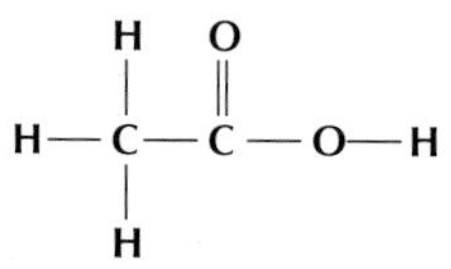   ethanoic acid

   (b)

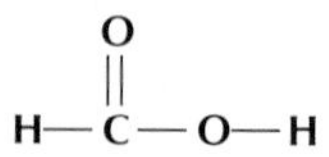

   methanoic acid 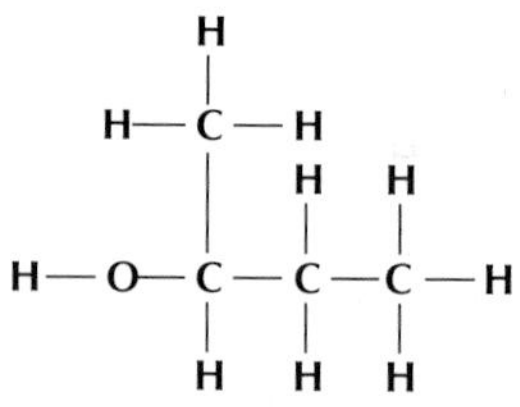   butan-2-ol

Quick Test 14

1. Glycerol has three hydroxyl groups in each molecule so can react with three fatty acid molecules.

2. Shake with bromine solution. The orange bromine rapidly loses its colour.

3. (a) In fats, the shape of the molecules allows them to pack closer together with stronger van der Waals forces between the molecules than is the case in oils. This means fats need more energy to separate the molecules, and are therefore solids at room temperature.

   (b) In a fridge the temperature is much lower than room temperature, so there is not enough energy to break the forces of attraction between the molecules, so the oil becomes solid.

Quick Test 15

1. (a) (i) and (ii)

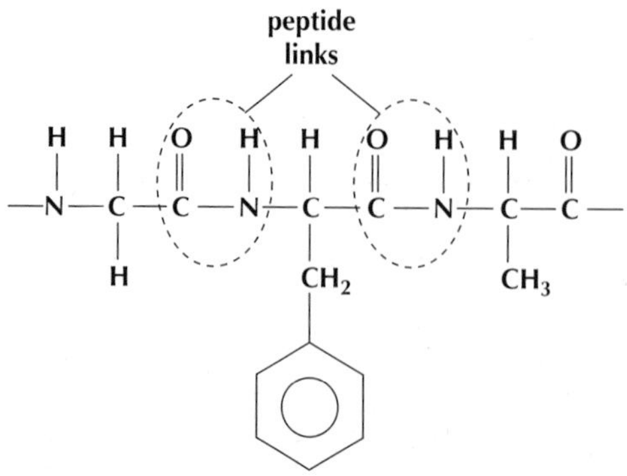

2.

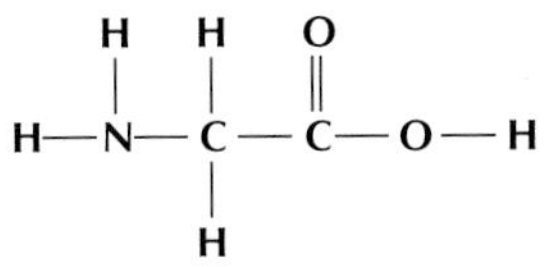

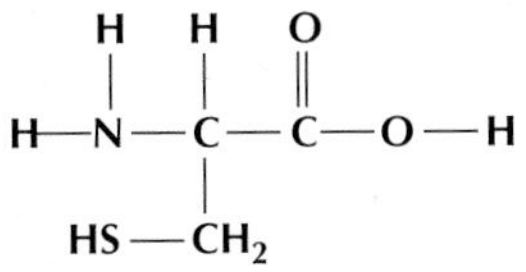

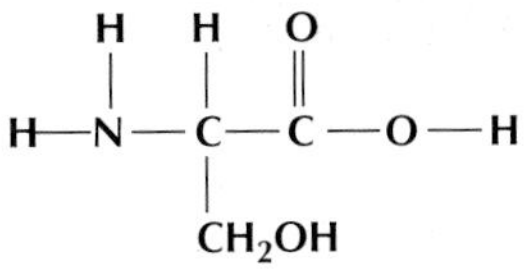

$$H-N(-H)-C(-H)(-H)-C(=O)-O-H$$

$$H-N(-H)-C(-H)(-CH_2-SH)-C(=O)-O-H$$

$$H-N(-H)-C(-H)(-CH_2OH)-C(=O)-O-H$$

(structures: three amino acids, with side groups H, $HS-CH_2$, and CH_2OH respectively)

Quick Test 16

1. (a)

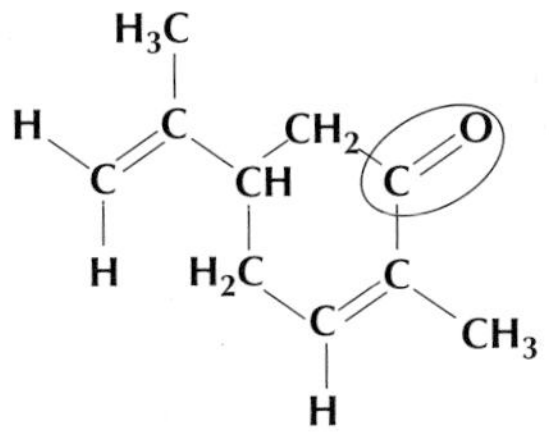

carvone

   (b) Ketone; the carbonyl group is not on the end carbon. The name ends in -one.

   (c) The smell of carvone (spearmint) is strong, which indicates it is volatile.

2. (a)

Aldehyde

pentanal

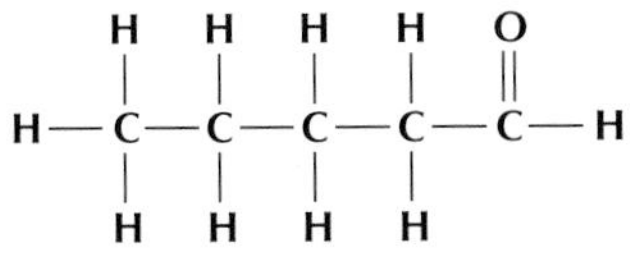

(structure, $C_5H_{10}O$)

Ketone

pentan-2-one   or   pentan-3-one

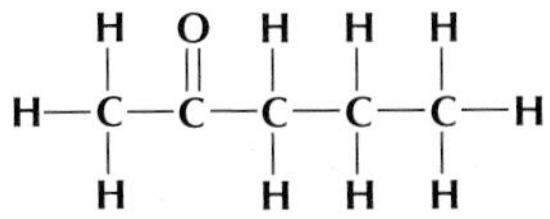

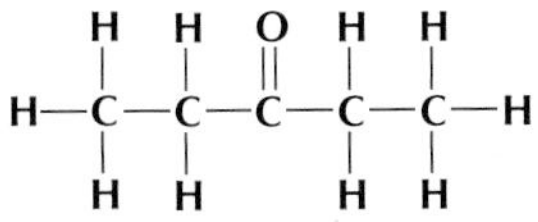

(structures, $C_5H_{10}O$)

   (b) Isomers

Quick Test 17

1. (a)

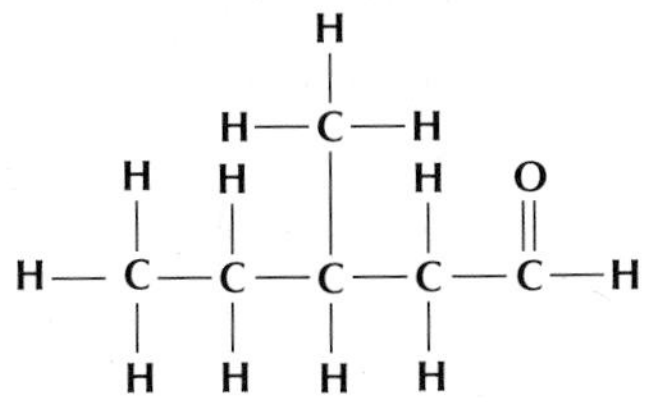

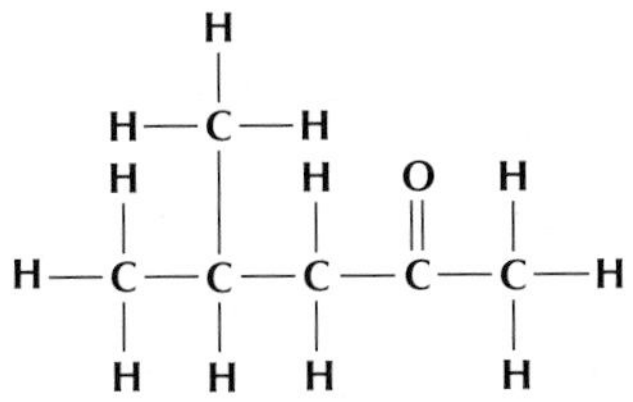

(two structures shown)

   (b) $C_6H_{12}O$

   (c) They have the same molecular formulae and different structures, and so are isomers. Isomers can belong to different homologous series.

2. (a) 4-ethyl-3-methylheptanal

   (b) 5,7-dimethyloctan-3-one

Answers to questions

Quick Test 18

1. (a) (i) acidified dichromate; (ii) green; (iii) ketone; (iv) Fehling's; (v) aldehyde.

   (b)

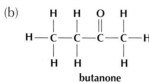

   butanone 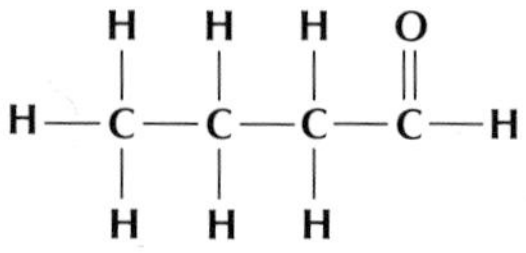   butanal

   (c) Carboxylic acid.

Quick Test 19

1. 5-ethyl-2-methyloctan-3-ol.

2.

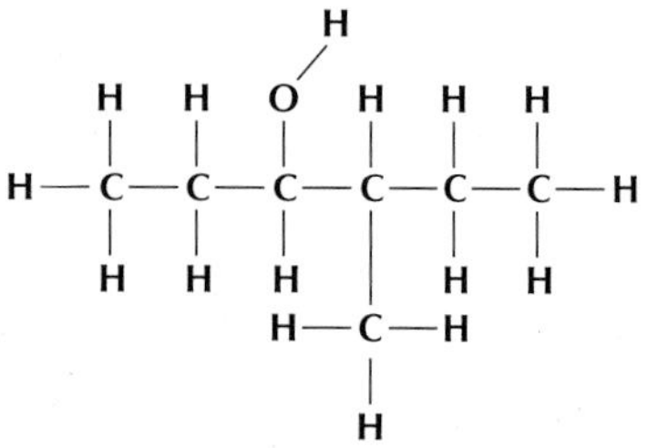

3. The molecules in fluoroethane are held together by weak London dispersion forces which do not need a lot of energy to separate them, so the boiling point is low. The molecules in ethanol have hydrogen bonding between them because of the –OH group on each molecule. Hydrogen bonds are stronger than London dispersion forces, so more energy is needed to separate the molecules, so the boiling point of ethanol is higher.

Quick Test 20

1. 2,2-dimethylpentan-1-ol: primary

2.

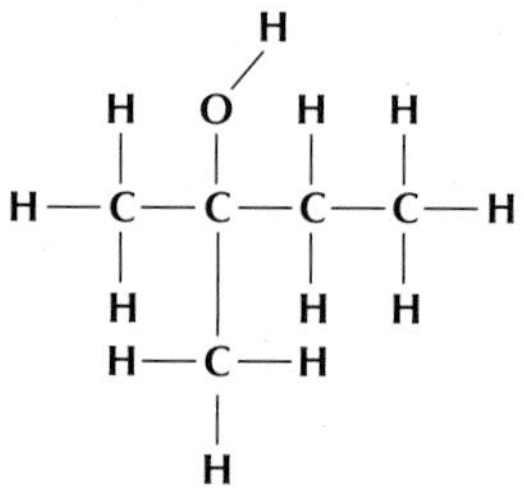

   tertiary

3. Alcohol in 1: Colour change from orange to green because 1. is a primary alcohol and is oxidised to an aldehyde.

   Alcohol in 2: No change because 2. is a tertiary alcohol and cannot be oxidised by acidified dichromate.

Quick Test 21

1. 2-ethyl-2,5-dimethylhexanoic acid

2.

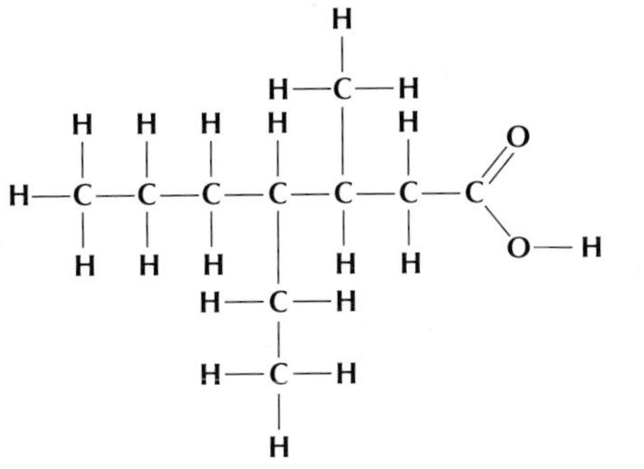

3. CH_3COOH $\rightarrow$ C_2H_5OH

   O:H   1:2 (2:4)   1:6

   The O:H ratio decreases so the reaction is reduction.

4. (a) $2C_2H_5COOH$ + $MgCO_3$ $\rightarrow$ $(C_2H_5COO)_2Mg$ + H_2O + CO_2

   (b) Magnesium propanoate

Quick Test 22

1.

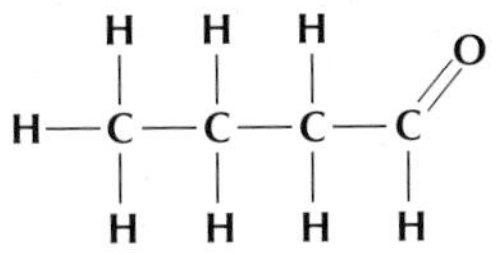

butanal 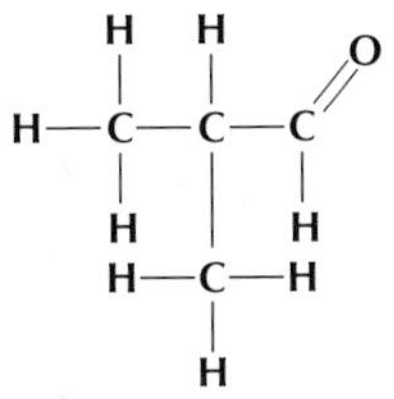   2-methyl propanal

2.

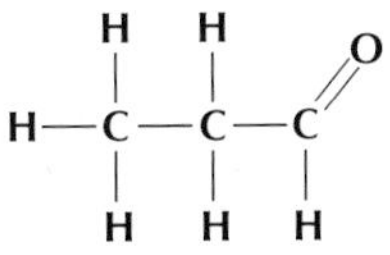

propanal 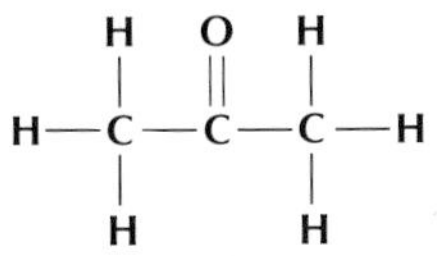   propanone

Quick Test 23

1. When a fat/oil reacts with a base, salts of the fatty acids form.

2. The hydrophobic tails of the soap ions burrow into the oil and grease. The agitation causes small grease droplets (micelles) to form in the water. The negative charges on the heads prevent the globules of oil from recombining. This allows the oil or grease to be washed off the surface.

3. Detergents do not form insoluble salts (scum).

Answers to questions

Quick Test 24

1. Emulsifiers are soap-like molecules. One or two fatty acid groups are attached to the glycerol backbone rather than the three in fats/oils. This results in the emulsifier having molecules with a hydrophobic part (the fatty acid 'tail') and a hydrophilic part – one or two hydroxyl groups or other water-soluble parts.

Quick Test 25

1. (a) Alcohols

   (b)
   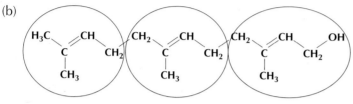

2. (a) Oxidation

   (b)

$$
\begin{array}{c}
CH_3 \\
\mid \\
C \\
\end{array}
$$

Quick Test 26

1. (a) (ii); (iv); (i); (iii)

   (b) Free radical chain reaction

   (c) (i) propagation; (ii) initiation; (iii) termination

   (d) UV light (radiation)

   (e) Breaking the C–Cl bond needs less energy.

Quick Test 27

1. Free radical scavengers are chemicals which react with free radicals that might otherwise cause damaging free radical chain reactions.

2. Any of the following (and also many more): vitamins C and E – foods and skin creams; polyphenols (flavonoids) – skin creams; benzophenone – plastics.

Quick Test 28

1. Waste prevention, more energy-efficient processes, reduce steps and by-products (improve the atom economy), use less hazardous and toxic chemicals, safer processes and products, design products which can be degraded in the environment, use more specific catalysts like enzymes, which reduce energy requirements.

Quick Test 29

1. (a) 30%; (b) 100%; (c) Although the percentage yield indicates low efficiency the atom economy is 100% so there are no useless by-products, which improves the overall efficiency.

2. (a) 79·8%; (b) 36·7%; (c) Although the percentage yield indicates high efficiency the atom economy is low because there are by-products, which reduces the overall efficiency.

3. £150.

Quick Test 30

1. 2·1 g
2. 1·36 g

Quick Test 31

1. 0·004 mol
2. 0·225 mol l^{-1}
3. 0·02 l (20cm^3)

Quick Test 32

1. (a) HCl by 0·003 mol
   (b) 1·22 g
2. (a) Cu by 0·03(05) mol
   (b) 13·49 g

Quick Test 33

2. (a) 3·56 l
   (b) 1·9 l
   (c) 9·86 g

Quick Test 34

1. (a) 500 cm^3
   (b) 350 cm^3
2. 1·49 l (0·0625 mol produced)

Answers to questions

Quick Test 35

1. (a) The reaction between hydrogen and iodine is reversible. As products are formed some of them re-form reactants, and an equilibrium position is reached where the rate of the forward and backward reactions is the same. At equilibrium the concentrations of reactants and products remain constant. The same equilibrium position is reached regardless of whether you start with reactants or products.

   (b) $H_2(g) + I_2(g) \rightleftharpoons 2HI(g)$

2. (a) (i) Right

   (ii) The concentration of products is greater than reactants at equilibrium.

   (b) At the point where the graphs level out.

Quick Test 36

1. (a) The equilibrium position moves left to form reactants because hydrogen ions (product) are being added, so the equilibrium moves to compensate for the imposed change.

   (b) The equilibrium position moves right. The hydroxide ions react with the hydrogen ions and so remove them. The equilibrium moves to compensate for the imposed change.

2. (a) (i) High pressure.

   (ii) An increase in pressure shifts the equilibrium in the direction of fewer moles of gaseous molecules, in this case to the right, so more methanol is produced.

   (b) When a product is removed the equilibrium shifts to the right to compensate for the change, i.e. more product is formed.

Quick Test 37

1. (a) Increased yield

   (b) The forward reaction is endothermic, so is favoured by an increase in temperature (the equilibrium shifts to the right).

2. (a) The equilibrium shifts to the left.

   (b) The forward reaction is exothermic, which means the reverse reaction is endothermic, which is favoured by an increase in temperature.

3. (a) The activation energy of the reverse reaction is lowered as well as the forward reaction.

   (b) The equilibrium position is not affected, but is reached more quickly than the uncatalysed reaction.

Quick Test 38

1. (a) The forward reaction is endothermic which is favoured by a rise in temperature. High temperature causes the equilibrium to move right.

   (b) High pressure would speed up the rate at which equilibrium is reached.

2. (a) Finely divided catalyst provides a bigger surface area on which a reaction can take place.

   (b) Removing the product causes the equilibrium position to move right, producing more ammonia.

   (c) The balanced equation shows that 1 mole of nitrogen reacts with 3 moles of hydrogen.

Quick Test 39

1. (a) $-1676 \cdot 4$ kJ mol^{-1}
   (b) A lot of heat produced when the alcohol burns is lost to the surroundings.
2. $19 \cdot 9°C$.

Quick Test 40

1. $-53 \cdot 5$ kJ mol^{-1}.
2. The same reaction is happening in each case: $H^+(aq) + OH^-(aq) \rightarrow H_2O(\ell)$

Quick Test 41

1. $+51$ kJ mol^{-1}.
2. -240 kJ mol^{-1}

Quick Test 42

1. -1680 kJ mol^{-1}
2. -1025 kJ mol^{-1}

Quick Test 43

	Oxidising agent	Reducing agent
(a)	$Cu^{2+}(aq)$	$Zn(s)$
(b)	$Br_2(aq)$	$2I^-(aq)$
(c)	$Ni^{2+}(aq)$	$Cr(s)$

Quick Test 44

1. (a) $Cl_2(aq) + 2e^- \rightarrow 2Cl^-(aq)$
   (b) $S_2O_3^{2-}(aq) + 5H_2O(\ell) \rightarrow 2SO_4^{2-}(aq) + 10H^+(aq) + 8e^-$
   (c) $4Cl_2(aq) + S_2O_3^{2-}(aq) + 5H_2O(\ell) \rightarrow 8Cl^-(aq) + 2SO_4^{2-}(aq) + 10H^+(aq)$
   (d) oxidising agent: $Cl_2(aq)$; reducing agent: $S_2O_3^{2-}(aq)$
2. (a) $S_4O_6^{2-}(aq) + 2e^- \rightarrow 2S_2O_3^{2-}(aq)$ reduction
   (b) $V^{3+}(aq) + 3H_2O(\ell) \rightarrow VO_3^-(aq) + 6H^+(aq) + 2e^-$

Quick Test 45

1. Strongest reducing agent: Li (lithium); strongest oxidising agent: F_2 (fluorine).

2. (a) (i) Yes; (ii) No; (iii) Yes; (iv) No

   (b) (i) $2I^-(aq) \rightarrow I_2(aq) + 2e^-$ oxidation

   $Br_2(aq) + 2e^- \rightarrow 2Br^-(aq)$ reduction

   (iii) $2Cl^-(aq) \rightarrow Cl_2(aq) + 2e^-$ oxidation

   $MnO_4^-(aq) + 8H^+(aq) + 5e^- \rightarrow Mn^{2+}(aq) + 4H_2O(\ell)$ reduction

   (c) (i) $2I^-(aq) + Br_2(aq) \rightarrow I_2(aq) + 2Br^-(aq)$

   (iii) $10Cl^-(aq) + 2MnO_4^-(aq) + 16H^+(aq) \rightarrow 5Cl_2(aq) + 2Mn^{2+}(aq) + 8H_2O(\ell)$

   (d) (i) Oxidising agent: $Br_2(aq)$; Reducing agent: $I^-(aq)$

   (ii) Oxidising agent: $MnO_4^-(aq)$; Reducing agent: $Cl^-(aq)$

Quick Test 46

1. (a) A

   (b) A is more soluble in the mobile phase so will travel further up the paper.

2. (a) It is very soluble in the stationary phase or not soluble in the mobile phase.

   (b) B and C. They have components which have travelled the same distance (have the same R_f value).

3. (a) 0·38

   (b) The R_f value is constant for a particular substance if other conditions such as paper and solvent used remain the same.

Quick Test 47

1. (a) If a pen were used the solvent might separate out the mixture which makes up the ink.

   (b) To make sure that substances which didn't separate with one solvent would with the other.

2. (a) The database is used to identify unknown substances from their R_f values. If the conditions are changed after the database is set up than the R_f values will not be accurate.

   (b) Caffeine. It has the largest area under the peak.

Quick Test 48

1. (a) 0·2 g

   (b) Weigh out accurately the required mass and dissolve it in deionised water, in a small beaker. Pour the solution into a volumetric flask, and rinse all of the solution out of the beaker using a water bottle. Add more deionised water until it reaches the etched line on the neck of the flask. Stopper the flask and invert it to ensure the solution is fully mixed.

2. 0·039 mol l^{-1}

3. 0·163 mol l^{-1}

Quick Test 49

1. Filtration.
2. Gas syringe.

Quick Test 50

1. $\dfrac{(17.7 + 17.6)}{2} = 17.65 \text{ cm}^3$

2. Student B